GROLIER
SCIENCE ACTIVITY SERIES

BIOLOGY

RESEARCH ACTIVITIES

GROLIER
SCIENCE ACTIVITY SERIES

BIOLOGY

RESEARCH ACTIVITIES

Barbara Newman

‹Vol. 1›

GROLIER EDUCATIONAL CORPORATION
Danbury, Connecticut

Edited by
Eugene Kutscher

© 1988 by

ALPHA PUBLISHING COMPANY, INC
Annapolis, Maryland

Produced in the United States of America by Alpha Publishing Co., Inc., 1910 Hidden Point Road, Annapolis, MD 21401

Printed in the United States of America

First Edition

SERIES ISBN: 0-7172-7130-7

Library of Congress Catalog Card Number : 88-82233

INTRODUCTION

GROLIER SCIENCE ACTIVITY SERIES

Biology Research Activities
Chemistry Research Activities
Earth Science Research Activities
Physics Research Activities
Preparation Guide Research Activities

The *Grolier Science Activity Series* is truly a remarkable series. It combines the rare but necessary ingredients to create a successful and rewarding learning experience. The series presents enrichment activities for the general science curriculum.

Each activity is self-contained and provides everything needed to gain a basic understanding of a concept or to work through a project. How one uses the activity depends only on ability and motivation. These ready–to–use projects can immediately be implemented and are especially ideal for independent study, rotating lab groups, remedial or makeup work.

Everything needed to proceed safely and successfully is included in the directions for each of the projects. The stimulating laboratory and investigative activities provide high motivation learning experiences for independent use.

An outstanding team of secondary science teachers has designed a practical supplemental tool. Each activity has been carefully prepared and tested by classroom teachers to ensure appeal to a broad range of student interest, ability levels and modes of learning.

Each book contains innovative and traditional projects for both the bright and average, the self-motivated and the those who find activity motivating. The field tested projects are exciting and provide practical experiences in the use of sound lab techniques and procedures through the application of the scientific method. Important vocabulary words have been bold faced and should be examined in a dictionary and learned.

The comprehensive *Preparation Guide* covers each activity in the four titles and discusses the goal, student objectives, prelab discussion, guide to the investigation, vocabulary or glossary, guide to additonal resources and suggestions for further study for each project.

ABOUT THE AUTHORS

This series was written by four members of the Roslyn, New York science department, nationally recognized for excellence. The school emphasizes scientific processes through an independent research program as well as in its classroom activities. Currently, eighty of the nine hundred students in grades nine to twelve are enrolled in the independent research program.

Sixteen students who were chosen to be honored by the Westinghouse Science Talent Search in the past three years for high school level research were from Roslyn High School. Numerous other grants and prizes were awarded as well.

The philosophy of educational excellence in science is an actuality at Roslyn, and the activities incorporated into these magnificent books reflect that standard. They are exciting, easy to comprehend, and yet lend themselves to concept development rather than to rote learning.

Biology Research Activities by Barbara Newman.

Ms Newman holds a B.S. in biology from Hofstra University and an M.S. in Science Education from Queens College of the City University of New York. During her nineteen years at Roslyn she has taught biology, chemistry, earth science, biochemistry, marine biology and laboratory technology, as well as serving as an advisor for students conducting research.

She has written curriculum guides for the laboratory technology, life science and self-paced biology courses. Ms. Newman has helped to develop experimental state syllabi, has written questions for the National Science Teachers Exam, and has presented workshops at the annual meeting of the Science Teachers Association of New York State.

Her broad experience was the basis for her selection to teach in the Roslyn School-Within-A-School program, which she did for many years.

Earth Science Research Activities by James Scannell.

Mr. Scannell holds an A.A.S. in Forestry from Paul Smith's College, a B.A. in Earth Science Education from Long Island University, and an M.A. in Science Education from SUNY, Stony Brook. During the past eight years at Roslyn, he has taught all levels of earth science, including eighth grade accelerated honors earth science.

In addition, Mr Scannell has taught astronomy and chemistry. He has been responsible for writing curriculum objectives and new laboratory activities for the Roslyn earth science program. Currently, he is completing a sound/slide show on Long Island's glacial history.

Chemistry Research Activities by David Williams Ph.D.

Dr. Williams is assistant to the Department Chairman at Roslyn High School. He holds a B.S. from M.I.T., an M.S. from the University of Vermont and a Ph.D. from the University of Florida, all in chemistry.

At Roslyn for eighteen years, Dr. Williams has taught all levels of chemistry, including advanced placement. He has taught physics, astronomy, biochemistry and physical science as well. In addition, he has written curriculum guides for both chemistry and biochemistry. He has advised over forty research students, six of whom have been named to the Westinghouse honors group.

Physics Research Activities by Eugene Kutscher.

Mr. Kutscher is Science Chairman and Coordinator of Science Research for the Roslyn Schools. He holds a B.A. and M.A. in Physics from the City University of New York. He has spoken on the topic of "How To Start A successful High School Science Research Program" at the New York State School Boards Association, the annual meeting of the Science Teachers Association of New York State, and under the sponsorship of the National Science Supervisors Association, at the annual conference of the National Science Teachers Association.

In addition, Mr. Kutscher lectures on nuclear and alternative energies, is active in Science Technology and Society groups, and is on the National Board of Directors of Zero Population Growth.

This year, he again conducted a National Science Supervisor Association endorsed workshop on high school science research and participated in a panel on the same subject at the National Science Teachers Association annual meeting. He has presided and spoken about population, energy and other science-technology-society issues at national and regional NSTA meetings, professional growth day workshops, on television's "Straight Talk" and at the Edison Electric Institute's meeting for utility educators.

Mr. Kutscher has been a science educator for twenty years, teaching physics, astronomy, biology, chemistry, earth science and mathematics in all grades from seven through college.

Table of Contents

WHAT IS ALIVE?

INTRODUCTION: Most of you have no problem recognizing a living animal or plant. What characteristics, however, contribute to this living condition? Must an organism move to be considered alive? Must it breathe? And how would you describe seeds? What processes must be occurring within the seed if it is to be considered alive? There is no one all–encompassing definition of **life**. And there are occasionally no "right" answers to these questions.

There are, however, many **life processes** that living organisms must carry out. **Inanimate**, or non–living objects, however, may also seem to carry out one or more of these processes. But do they possess all the characteristics of living systems? **Transport** of materials, **growth, respiration, regulation, nutrition, excretion,** and most importantly, **reproduction**, are all processes considered essential to life.

In this exploration you will investigate many objects, some alive, some inanimate, and some dead. Through careful observation you are going to attempt to classify each specimen. In the process, you will make your own conclusions regarding the definition of life.

PURPOSE:

- What are the characteristics of living organisms?

MATERIALS:

Various samples as selected by your teacher.

PROCEDURE:

Your teacher may assign you to work individually, or in groups, for this activity.

Carefully observe the various samples available. Remember, scientific observation often includes more than just "looking." In your data section, describe any characteristics of each sample that can help you classify it as living, dead, or non–living.

Life processes are not always easy to observe. In some instances you may make **inferences** about a specimen, without actual evidence. If your assumptions regarding a particular sample are not supported by direct observational data, be sure to list them as inferences.

After completing your observations, and any discussion suggested by your teacher, attempt to classify each of your specimens as living, dead or inanimate.

DATA:

Sample	Observations	Inferences	Conclusion (Status)
1			
2			
3			
4			
5			
6			
7			
8			

Add more if needed

CONCLUSIONS:

1. Describe the processes considered to be essential to living organisms.

2. Why is it sometimes difficult to determine if something is alive?

3. In your **own words,** write a definition of "alive."

SUGGESTIONS FOR FURTHER STUDY:

- You are a scientist on the first manned space flight to Saturn. You discover many "jelly fish–like things" floating in the thick atmosphere. It's your job to determine if these "things" are living organisms. What would you look for? What tests could you perform? What data would you need to make a valid, scientific, decision?

- There are several aspects of science/society in which the question of life status is central. A few of these include the nature of viruses, human embryos (or even frozen embryos) and coma patients on life support systems. Select a specific area of controversy and investigate the scientific evidence that either supports or contradicts the "life" criteria you have established. Come to your own conclusion, and support this with documentation.

- Investigate and report on the Martian soil sample experiment conducted to determine if there is life on Mars.

SCIENTIFIC MEASUREMENTS

INTRODUCTION: Scientific measurements are always made in the **metric system.** The United States is just about the only country in the world that does not use this system regularly. Not only is the metric system more universal than our English system, it's also much easier to use. Since it is a decimal system, there are never any fractions to deal with. Also, each smaller unit is related to a larger one by one or more powers of ten, so that converting from one unit to another merely requires moving a decimal point. Length, volume and mass are measured in the units **meters, liters** and **grams,** respectively. Temperature is measured in **degrees, Celsius (centigrade).**

If you are not yet familiar with the divisions commonly used in the metric system, refer to the chart below, which summarizes some of the more commonly used units.

	PREFIX		UNIT	PREFIX
MICRO–	MILLI–	CENTI–	GRAM or METER or LITER	KILO–
(.000001)	(.001)	(.01)		(1000 X)
SMALLER ←				→ LARGER

How **accurately** you can measure something depends upon the instrument you use. The **precision** of your measurement refers to its reliability, or how easily your measurement can be reproduced. Because of the limits in the accuracy of your instruments, there will always be some uncertainty in your measurements. For example, if you had a metric ruler that was divided into millimeter units, you could only be **certain** of a measurement to the nearest millimeter. However, it is customary to include one extra digit, an **estimated** number, that goes beyond the exact units that are known. Look at the ruler below:

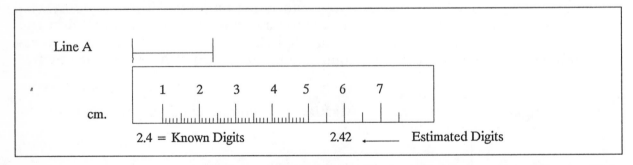

Line A

cm.

1   2   3   4   5   6   7

2.4 = Known Digits   2.42 ____ Estimated Digits

To indicate the length of line A, you can say with certainty that it is 2.4 cm. You should, however, include one more digit, the estimated one, in order to report your measurement most accurately. In this instance then, you might record the length as 2.42 cm, realizing that the last digit is actually an estimate. In scientific measurement, all the digits known with certainty, plus one estimated digit, are known as **significant figures.** Significant figures are important in science, since you always want to give as much accurate information as possible, but never more than you can truthfully report.

PURPOSE:

- To make scientifically accurate metric measurements of length, volume, mass and temperature.

MATERIALS:

Metric ruler
Glass slide
Graduated cylinder
1 ml pipette
Small rubber stopper
Test tube
Balance

Ice bath
Boiling water bath
Celsius thermometer
Goggles
Small beaker
Penny

PROCEDURE:

Part I: LENGTH

1. Obtain a metric ruler, and use it to measure the width of this page, in centimeters. Record your answer, to the proper number of significant figures, on Data Table 1.

2. By moving the decimal place forward or backward, convert the measurement you just made into millimeters and kilometers.

3. Measure the length, width and height of a glass microscope slide. Record these measurements, to the proper number of significant figures, in Data Table 2. Be sure to include the proper units.

Part II: VOLUME

1. A **graduated cylinder** is commonly used to measure volume. Obtain a 50 ml graduate. Look at the markings on the cylinder. What is the most accurate *known* measurement you can make using this piece of equipment?

2. When reading the level of a liquid using the graduate cylinder, you must read the volume at the *center* of the **meniscus.** (See illustration). Be sure to take your readings with the liquid at eye level.

3. Practice using the graduate cylinder by attempting to fill the cylinder with exactly 18.50 ml of water. Now try to remove exactly 5.50 ml of water.

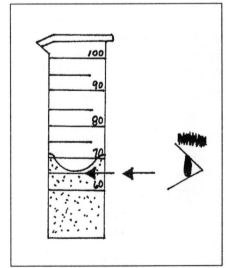

Reading a graduate cyliner

4. The volume of a solid object may be determined by calculating the amount of water the object displaces. Determine the volume of a rubber stopper by the following procedure: (Record your data in Data Table 3).

   a. Fill a graduate with 20.00 ml of water.

   b. Immerse the rubber stopper in the cylinder.

   c. Read the new water level in the graduate.

   d. Subtract 20.00 from the new water level to determine the volume of the rubber stopper.

5. Repeat the **water displacement method** to determine the volume of a penny. Be sure to include proper units and significant figures.

6. A **pipette** is another piece of equipment commonly used to measure (or transfer) small amounts of liquid. A pipette works like a straw. You can draw up a certain volume of water into the pipette, cover it with your index finger, and release the liquid into another vessel. Some pipettes are equipped with a suction device, so that the experimenter does not have to put the pipette into his/her mouth, or be concerned about toxic or caustic liquids accidentally being swallowed.

7. Obtain a sterile, 1 ml pipette. What are the smallest known units that can be measured with this device? Place some water into a clean beaker. Practice using the pipette by attempting to transfer 0.80 ml of water from the beaker into a test tube. If you have not used a pipette before, this technique will take some practice, so be patient.

8. Which piece of equipment is more accurate, a 1 ml pipette or a 50 ml graduate cylinder? Explain.

Part III: MASS

1. Mass is measured in the science laboratory with a **balance**. Various types of balances are available, but they are all based on the same principle: comparing an experimental mass with a known mass so that the scale of the balance rests at zero.

2. Look at the balance available to you. What type of balance is it? What is the minimum and maximum mass that can be measured with this instrument? What is the smallest known unit on your balance?

3. Measure the mass of a rubber stopper and a penny (from Part II). Record your data in Data Table 4. Be sure to use significant figures and proper units labels.

4. How could you determine the mass of a liquid using the balance? Devise a method for calculating the mass of 150 ml of water. On the Data Sheet, describe your procedure. Perform your measurements and record these in an appropriate fashion.

Part IV: TEMPERATURE

1. Obtain a Celsius thermometer. Look at its markings. What is the smallest known unit on the instrument?

2. Determine the freezing point of water by inserting the thermometer into a beaker of ice, and reading the temperature, in significant figures, when the mercury stabilizes. Record your data in Data Table 5.

3. Determine the boiling temperature of water. **Wear goggles when boiling water.** Record this data.

4. Determine the temperature of your classroom. Record the data.

Part V: PERCENT ERROR

1. The precision with which you made your measurements can be determined by comparing your measurements with KNOWN measurements for the same object. Your **percent error** indicates how much you are "off" in your measurements. Percent error is calculated according to the following formula:

$$\% \text{ error} = \frac{\text{difference between measured \& actual values}}{\text{actual value}} \text{ X } 100$$

2. Ask your teacher for the actual values for the measurements you made in this investigation. You will need the actual values for the following data:

    Width of the paper
    Length, width and height of the glass slide
    Volume and mass of the rubber stopper
    Volume and mass of the penny
    Freezing, boiling and room temperatures

Write the given data in Data Table 6.

3. Determine your percent error for each of these measurements, and record the results in the data table.

4. For which measurement did you have the greatest percent error? Why do you think this was so?

5. For which measurement did you have the least percent error? Why?

DATA:

Part I

DATA TABLE 1: WIDTH OF PAGE

cm.	mm.	km.

DATA TABLE 2: MEASUREMENT OF GLASSS

Length	Width	Height

Part II

1. _____

DATA TABLE 3: WATER DISPLACEMENT MEASUREMENT OF VOLUME

Object	Initial Volume	Volume with Object	Volume of Object
Rubber Stopper	20.00 ml		
Penny	20.00 ml		

7. _____

8. _____

Part III

2. _____

_____

DATA TABLE 4: MASS

Object	Mass (g)
Stopper	
Penny	

4.

Part IV

1._____

DATA TABLE 5: TEMPERATURES °C

Freezing H$_2$O	Boiling H$_2$O	Room Temperature

Part V

DATA TABLE 6: PERCENT ERROR OF MEASUREMENT

Object Measured	Experimental Measurement	Actual Measurement	% Error
Width of Paper			
Length of Slide			
Width of Slide			
Height of Slide			
Volume of Stopper			
Volume of Penny			
Mass of Stopper			
Mass of Penny			
Temp. of Ice			
Temp. of Boiling H$_2$O			
Room Temp.			

4._____

5._____

CONCLUSIONS:

1. Fill in the blank spaces below to demonstrate your knowledge of the metric system:

   a. 55 cm =_____ mm =_____ m

   b. 0.75 ml =_____ L

   c. 9.02 g =_____ mg =_____ kg

2. What is meant by *significant figures* in measurement?

3. Three students, A, B, and C, measured the size of a seedling, using a standard metric ruler. Their measurements were 2.255 cm, 25 mm, and 2.18 cm, respectively. Which student took the most accurate measurement, and why?

4. What are the advantages of the metric system for measurement?

5. Distinguish between accuracy and precision in measurement.

SUGGESTIONS FOR FURTHER STUDY:

- Using a chemistry or physics book for reference, describe the rules for using significant figures in calculations. Explain when zeros do, and do not, count as significant digits. Demonstrate your understanding of these concepts by calculating the area of this page, in significant metric units.

- If an analytical balance is available to you, find out how it works, and explain its use. Use the analytical balance to measure the mass of the rubber stopper and penny that you used in this investigation. How does the accuracy of these measurements compare with those you obtained from the traditional balance?

THE COMPOUND MICROSCOPE: PART I
INVESTIGATING THE INVISIBLE WORLD

INTRODUCTION: One of the most basic instruments used in the study of biology is the **compound microscope.** Since you will undoubtedly be using this piece of equipment throughout the year, it is essential that you become familiar and comfortable with it. With a little instruction, and a lot of practice, a whole new microscopic world will be opened to you.

The primary functions of the compound microscope are **magnification** and **resolution.** As light passes through a specimen placed on the stage, the rays are eventually spread out by a system of glass lenses. By looking through the final lens, you not only see the enlarged specimen, but you are also able to distinguish separate images that you would ordinarily see blended together.

PURPOSE:

- How can the compound microscope be used effectively?

MATERIALS:

Compound microscope
Lens paper
Glass slide
Cover slips

Medicine dropper
Small beaker of water
Lower case newsprint letter

PROCEDURE:

Part I: STRUCTURE AND CARE OF THE MICROSCOPE

a. The compound microscope is an expensive and delicate instrument. It is your responsibility to care for it properly. Obtain a microscope from your teacher. Always carry the microscope with two hands; one under the base, and the other on the arm (see illustration 1). Place the microscope gently on your desk, at least five centimeters away from the edge.

b. Compare your microscope to the one shown in illustration 2. Since there are many types of compound microscopes available, yours may differ in one or more ways from the one pictured here.

> 1. On your data sheet, describe any basic differences between your microscope and the one in illustration 2.

c. You are now going to become more familiar with the care and operation of the various parts of the microscope. Always refer to *your* microscope, since it may differ from the one in the illustration.

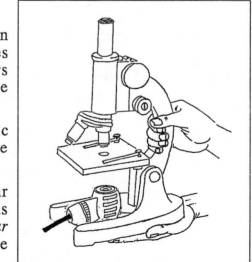

illus. 1

Wipe all the lenses (**objectives and ocular**) with lens paper. Never use other materials to clean the lenses, since they scratch easily. Get in the habit of cleaning the lenses whenever you use the microscope.

Find your light source. If your microscope has a built–in **illuminator**, turn it on. If your microscope is equipped with a **mirror**, adjust it so that the light is reflected up through the stage.

The amount of light reaching the specimen is critical, and may be controlled by adjusting the **diaphragm**. Locate the diaphragm underneath the stage of your microscope. While looking through the ocular (**eyepiece**), adjust the diaphragm so that the maximum amount of light is reaching your eye. When you are looking at actual specimens, you may

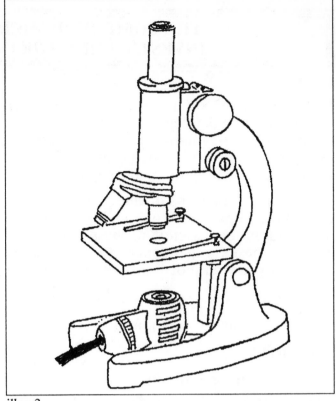

illus. 2

not always need maximum illumination. Practice adjusting the diaphragm.

The objective lenses are those closest to the specimen you will be viewing.

2. How many objective lenses does your microscope have? What is the magnifying power of each of your lenses? (Look on the barrel of each lens).

The objective lenses are attached to a **revolving nosepiece** so that you may switch from one power to another by simply turning the nosepiece. Rotate the nosepiece until you hear a click, indicating the objective is in its proper place.

3. Aside from the magnification indicated on the lens, how can you distinguish one lens from another?

The microscope can be focused on a particular specimen by moving the **coarse** or **fine adjusters**. Locate these knobs on your microscope. The coarse adjuster is the larger of the two. Slowly turn the coarse adjuster all the way in one direction, then the other

4. Which part of your microscope moves? Do the same with the fine adjuster. Do you notice any movement? What do you think is the difference between the two adjusters?

Other parts of the microscope include the stage, the base, the arm, and the stage clips. Look at the microscope and determine the function of these parts. Now look carefully at the ocular. You should see a number printed on it.

5. What does this number signify?

Part II: PREPARING AND OBSERVING A WET MOUNT SLIDE

a. A **wet mount** is a temporary slide that you prepare in order to view a specimen under the microscope. In this exercise you will prepare a wet mount of a letter of newsprint. Find a **small, lower case** letter and cut out a small square around it. Do **not** select the letters c, i, s, o, l, t or x. Place your letter on the center of a clean glass slide. (If your slide is dirty, wash it in soap and water and blot dry with lens paper). See illustration 3.

b. Using a medicine dropper, place a drop of water over your specimen, being careful not to touch the end of the dropper to the newsprint. If your letter moved when the water was added, use a dissecting needle to straighten it out.

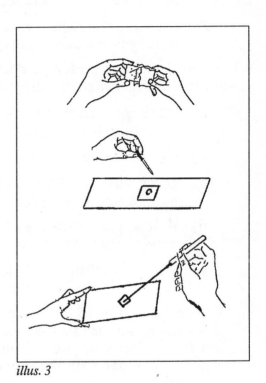

c. Now you are going to place a **cover slip** over your specimen. If the coverslip is smudged, wipe it gently with a folded piece of lens paper. In order to prevent air bubbles from being trapped under the coverslip, lower the coverslip on an angle, using the dissecting needle. Your finished slide should be free of air bubbles, with the cover slip floating on the surface of the water drop.

d. With your microscope in an upright position, place your slide on the stage, **with your letter in its proper, upright position**. Adjust the slide so that your letter is directly in the center of the stage, over the opening. Secure the slide with the stage clips.

illus. 3

e. Click the lowest power objective lens into place. While looking through the eyepiece, adjust the diaphragm to allow in as much light as possible. Now, while watching the lens, turn the coarse adjustment knob until the lens is as close to the slide as possible. Some microscopes have an automatic stop, so that you can not crack the slide with the lens. If your microscope does not have this feature, be careful that you do not lower the lens too much. While again looking through the ocular, turn the coarse adjuster slowly until you have focused on your specimen. If you try to focus too quickly, you may go right past the spot of focus and have difficulty finding the object.

f. Once you have the specimen in focus, turn the fine adjuster slowly to make the image as clear as possible. Reevaluate your light conditions by turning the diaphragm to the setting that lets in the amount of light you would most prefer.

g. Now you should be looking at your newsprint letter under low power magnification.

6. What is the total magnification of your image? (Be sure to multiply the magnification of the ocular by that of the objective lens you are using). Other than its increased size, in what ways does your image differ from the letter as seen on your slide?

h. While looking through the eyepiece, move the slide to your right.

7. What do you see through the microscope? What do you see when you move the slide to the left? Up? Down?

8. On a separate sheet of white, unlined paper, draw a sketch of your letter as it appears under the microscope.

Whenever drawing a sketch of a microscopic image, be sure to follow the following guidelines, unless otherwise directed by your teacher:

a) Make your drawings large, but simple. Have your drawing take up at least one–half a page of unlined paper.

b) Use a #2 pencil.

c) Draw **only** what you actually see through the microscope. Include any details that you do see.

d) Below your drawing, indicate the total magnification used. (For example, 40x). Above, write the title of your sketch.

e) When labeling parts of your drawing (not needed in this exercise), use straight label–lines, and write all words horizontally, if possible.

i. Before switching to a higher magnification, be sure your image is centered. Turn the nosepiece to the next highest objective lens (If your microscope only has two objectives, you will now be on the highest power). When focusing on high powers, **use only the fine adjuster.**

9. Why is this an essential procedure?

10. Is your image lighter or darker under the higher magnification? Why do you think this is so?

j. Readjust the diaphragm until your light is satisfactory.

11. How does the size of the microscopic field (the circle of light you see through the eyepiece) differ under a higher magnification? How much of your specimen can you see now? Why?

12. Make a sketch of your letter as it appears under high power. Be sure to draw only what you see. Indicate the total magnification under your sketch.

DATA: Part I

1. _____

_____

2. _____

_____

3. _____

_____

4. _____

_____

5. _____

_____

Part II

6. _____

_____

7. _____

_____

9. _____

_____

10. _____

_____

11. _____

_____

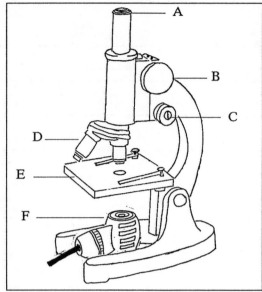

illus. 4

CONCLUSIONS:

1. Label the parts indicated on the compound microscope (illustration 4). Briefly describe the function of each labeled part.

NAME OF PART   FUNCTION

A. _____   _____

B. _____   _____

C. _____   _____

D. _____   _____

E. _____   _____

F. _____   _____

2. Question? (Fill in the chart below)

Magnification	Relative size of field	Relative bright-ness of field	Total magnification
Low			
High			

2.   Complete the chart below, which compares various aspects of high and low powers of the microscope.

3.   A student was viewing a specimen at low power, as shown in the illustration on the right. What would the student see if he/she switched to high power? In what direction(s) should the student move the slide in order to center the image?

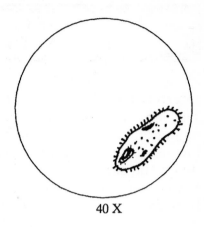

40 X

SUGGESTIONS FOR FURTHER STUDY:

*   Other types of microscopes are often used in biological studies. One of these, the steriomicroscope, is often available in schools. If you have such a microscope available, compare its capabilities to those of the compound microscope. Ask your teacher for specimens to view under the steriomicroscope, or select appropriate objects yourself. Write a report, including diagrams, comparing and contrasting the two types of microscopes.

*   Pond water is usually teeming with various microscopic organisms; both plant and animal. Get one or more samples of pond water and prepare several slides for examination. Using a reference table to help you identify what you find, sketch any organisms you find. Write–up your investigation, including statements about the relative numbers of the various organisms you find.

*   The electron microscope, or scanning electron microscope, has become an invaluable tool to the modern scientist. Research the structure and uses of this type of microscope. Compare and contrast it to the optical microscope you have been using. If you live near a large hospital or university, try to set–up a visit to the electron microscopy unit.

THE COMPOUND MICROSCOPE: PART II
RESOLVING POWER , DEPTH OF FIELD AND MEASUREMENT

INTRODUCTION: Even tiny objects viewed under the microscope have depth. The lenses of the microscope allow you to see the difference between a "nearer" object and a "farther" object more easily than you could with your own eyes. With low power, however, two objects (such as crossed strands of thread) might seem to lie on the same plane, i.e., they may both look in focus at the same time. At higher power, one of the threads should be clearly in focus "on top" of the other thread. In order to focus on one of the threads clearly, the underlying threads will be out of focus. High power shows less **depth of field** than does low power.

Resolving power is the ability to distinguish between two separate points that are very close together. It is the resolving power of the microscope that enables you to see a crisp, clear image. You have already seen evidence of the resolving power of your microscope when you looked at the letter of newsprint. With your eyes, you could not see all the little bumps and dots that actually make–up the newsprint. In this section, you will see further evidence of the resolving power of your microscope.

When measuring objects seen under the microscope, a new, smaller unit of measurement must be introduced. The common unit used for microscopic studies is called the **micrometer**. One micrometer is equal to .001 mm. In other words, **1mm = 1,000 u.** Unless your microscope has a built in measuring device (either in the ocular or on a mechanical stage), you will need to determine the diameter of your microscopic field before you can estimate the size of objects viewed under it. Once the diameter of your field of view is known, you will be able to estimate the size of objects within that field.

PURPOSE:

- What is meant by resolving power and depth of field?

- How can the size of microscopic objects be estimated?

MATERIALS:

Compound microscope	Lens paper
Glass slides	Color section from a magazine
Cover slips	Colored threads or hair
Medicine dropper	Clear metric ruler
Small beaker of water	Prepared slides (see teacher)

PROCEDURE: Part I: Depth of Field

a. To illustrate the phenomenon of depth of field, prepare a wet mount of three different color threads, crossed in the center of the slide. (You can also perform this activity using strands of different colored hair). Focus your slide under low power, concentrating on the area where the threads cross each other.

1. Can you tell the order of the threads? Are they all in focus at the same time?

b. Now switch to a higher magnification. Try focusing on each thread individually. Either begin from the "lowest" thread and focus up to each of the next threads, or focus "down" from the top thread.

2. Describe your results on the data sheet. Could you get all three threads in focus at the same time under high magnification? Why or why not?

Part II: Resolving Power

a. Now prepare another wet mount of a small square of colored magazine print. Focus on one of the colors under low power.

3. Describe what you see on your data sheet.

4. What does the resolving power of the microscope tell you about the way in which color is printed in a magazine?

b. Switch to high power.

5. Is the resolving power better under high or low power? (Hint – under which power is the image the clearest?)

Part III: Measuring Under the Microscope

a. You must first determine the diameter of your low power field of view. In order to do this, place a metric ruler on the stage of your microscope and focus on the edge of the ruler under low power. You should see an image similar to the one in the illustration.

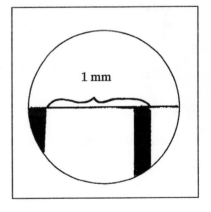

1 mm

b. In order to determine the diameter of your field, count the spaces between each millimeter line. Be sure to include any partial spaces at either end of the field.

6. Indicate the diameter, in millimeters, and the magnification, of your low power field. (Estimate your diameter to the nearest tenth of a millimeter.)

7. Convert your diameter measurements to micrometers according to the following equation: **micrometers = 1,000 x millimeters**

c. Obtain a prepared slide from your teacher. Printed on the edge of the slide is the name of the organism, or part of an organism, found on the slide. Focus this specimen under low power of your microscope. Remember to adjust your light for best viewing.

d. Estimate the length, or diameter, of an individual organism as seen through the microscope. To accomplish this, you must determine how many of your specimens could fit across the diameter of the low power field. If your prepared slide shows many cells filling up the field of view, you can simply count how many cover the diameter of the field. If, however, your slide shows scattered individual specimens, you have to **estimate** how many of these would fit, end to end, across the diameter of the field.

   8. On your data sheet, write the name of your specimen and how many can fit across the field of view.

   9. Determine the size, in micrometers, of an individual specimen. This can be found by dividing the diameter (in micrometers) of the field by the number of organisms that fit across it.

e. If you wanted to estimate the size of a specimen under the high power of your microscope, you would first have to determine the size of your high power field of view. This can not be done in the same manner as before, however. To see why, try to focus on the edge of your ruler when looking under high power.

   10. Can you see the millimeter markings under high power magnification? Why not?

f. The high power field diameter can be determined by calculating the ratio of the high power magnification compared to that of the low power. The diameter of the field of view is **inversely proportional** to the differences in magnification. That is, if the magnification of your high power field is ten times greater than that of your low power, then the diameter of the field of view under high power is ten times **smaller** than the field under low power. A simple equation that can be used to calculate the diameter of your high power field is:

$$\frac{\text{high power magnification}}{\text{low power magnification}} = \frac{\text{low power diameter (u)}}{\text{high power diameter (u)}}$$

After cross multiplying to solve for high power diameter, the equation results in:

$$\text{high power diameter(u)} = \frac{\text{low power magnification X low power diameter(u)}}{\text{high power magnification}}$$

   11. Determine the diameter of your field of view under high power magnification. (Be sure to indicate the magnification of your high power lens).

g. Focus on your prepared slide again, under high power magnification.

   12. How many specimens can fit across your high power field of view? What is the size of your specimen as calculated under high power magnification?

h. Since your specimen has not really changed its size just because you switched objectives, the measurements of its size should be the same when calculated under high and low powers. However, you probably obtained different figures for the size of your specimen when measured under the two different magnifications.

   13. Why do you think you calculated different sizes under the two magnifications? Which of your measurements is more accurate? Why?

DATA: Part I

1. _____

2. _____

   _____

Part II

3. _____

   _____

4. _____

   _____

5. _____

   _____

Part III

6. _____

7. _____

8. _____

9. _____

10. _____

11. _____

12. _____

13. _____

   _____

CONCLUSIONS:

1. In what ways can the concept of depth of field be important when you are viewing a three dimensional specimen, such as a cell?

2. Which is better, the resolving power of your eyes or that of the compound microscope? How can you tell?

3. If a single–celled organism swam at 300 micrometers per second, how long would it take to cross the low power field of your microscope?

4. Find the diameter of the high power field of a microscope that has a low power objective of 10X, a high power objective of 50X, and a low power field of view of 1500 micrometers.

SUGGESTIONS FOR FURTHER STUDY:

*   Determine the difference in **area** of your high and low power fields of view. When would this information be important in microscopic work?

*   Estimate the sizes of various microorganisms you find in a sample of pond water. Sketch each, and try to identify them with the help of a field guide book or reference chart for fresh water Protozoans.

*   Estimate the average diameter of human hairs of different colors and degrees of curliness.

HOW IS THE STRUCTURE OF A CELL RELATED TO ITS FUNCTION

INTRODUCTION: Part of the **cell theory** states that cells are the basic units of structure and function in living organisms. In both plants and animals, cells share many common structures and **organelles**. Even within the same kingdom there may be significant differences in structure among cells with different functions. Between the kingdoms, these differences may be even more obvious. In this investigation, you will examine various plant and animal cells, relating their differences in structure to their different functions.

PURPOSE:

- How is the structure of a cell related to its function?

MATERIALS:

Microscope	Elodea leaf
Piece of onion	Cover slips
Prepared slides:	Glass slides
Red blood cells	Lugol's iodine
Sperm	Toothpicks
Neuron	Single–edge razor blade
Lens paper	Forceps

PROCEDURE:

Part I: PLANT CELLS

1. Obtain a microscope, slide, cover slips, lens paper and a dropper bottle of iodine. Be sure all your materials are clean before proceeding.

2. Get a slide ready by placing a small drop of iodine in the center. Put the slide on the side of your desk. **Take care when handling the iodine. It is a stain, and will stain your fingers as well as the cell.**

3. Remove a thin, small piece of onion epidermis. This is most easily done by snapping off a small piece of the interior of an onion section, then peeling off a single layer with your forceps. See illustration 1.

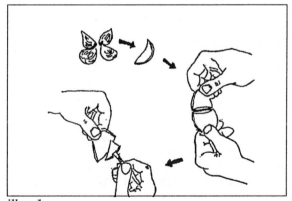
illus. 1

4. Place your section of onion epidermis in the drop of **stain** on the slide you just prepared. If the onion becomes wrinkled, smooth it out using dissecting needles. Lower a cover slip over your specimen, and blot any excess stain with paper towel.

5. Observe your slide under low power. Describe the general appearance of the onion cells on your data sheet. What function(s) could these cells serve?

6. Switch to the next highest magnification. Readjust your light if needed.

7. Draw a picture of one onion cell on a separate piece of plain, unlined paper. Remember to make your sketch large and accurate. Use a #2 pencil, and avoid shading. On your drawing, label the following organelles: **cell wall, cytoplasm, nucleus.**

8. Using the fine adjuster, look carefully for "empty" spaces within the cytoplasm. If you see any of these **vacuoles**, add them to your sketch. Also look carefully at the inside edge of the cell wall. Can you see the **plasma (cell) membrane** that encloses the cytoplasm? If so, add this to your drawing as well.

9. Remove your slide and wash it. Now prepare a slide of an Elodea leaf. Again, put a small drop of iodine in the center of your clean slide. With forceps, remove one leaf from the Elodea sprig and place it in the iodine on your slide (see illustration 2.) Add a cover slip and blot away excess stain.

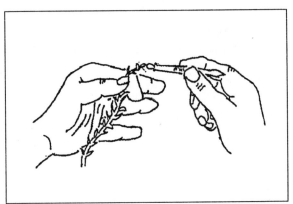

illus. 2

10. Observe your slide under low power. Compare the general shape of the Elodea cells to those of the onion. Record any similarities/differences on your data sheet.

11. Switch to high power magnification. Adjust your light for best visibility. On the same paper on which you drew the onion cell, draw one Elodea cell. (Don't forget to title each drawing and indicate the magnification used). Label the following parts of the Elodea cell: **cell wall, cytoplasm, nucleus, chloroplast.**

12. Look carefully and try to see the **plasma membrane** and **vacuole**. If visible, add these to your drawing.

13. Remove and clean your slide.

Part II: ANIMAL CELLS

1. Prepare a slide of human **epithelial** cells, using cells from your cheek. First, get a clean slide ready by placing a drop of iodine in the center. To obtain epithelial cells, *gently* scrape the inside of your cheek with the end of a toothpick. Smear the end of the toothpick in the iodine on your slide. Lower a cover slip and blot away any excess stain with a paper towel.

2. Observe your slide under low power. Look for cells that are isolated as opposed to those you may see in clumps. Center a cell in the field of view and switch to high power.

3. How does the shape of your cheek cell compare to the general shape of the plant cells you saw? What structure, missing in the cheek cell, appears to be responsible for the rigidity of the plant cells?

4. Make a sketch of one cheek cell. Label the following: **cell membrane, nucleus, cytoplasm.**

5. Using the fine adjuster, decide whether the cheek cell is considered thick or thin. How does its thickness compare with the plant cell's?

6. Do you see any large vacuoles like the ones you saw in the plant cells? What may be the significance of this

7. Remove your slide and clean it.

8. Obtain a prepared slide of nerve cells, or **neurons.** Clean the slide and view it first under low, then high power.

9. How does the general shape of these cells compare to all the other cells you have seen? What is the most obvious difference between neurons and cheek cells?

10. Make a sketch of a single neuron. Label the **cell body** (containing the cytoplasm), **nucleus**, and the extensions, called **dendrites.**

11. Obtain a prepared slide of blood. After cleaning it, focus first under low, and then high power.

12. The most numerous cells on the slide are the red blood cells. How does their shape compare to the others? List any cell organelles you can see.

13. Make a sketch of a single red blood cell. Label the **cell membrane** and **pigment.**

14. Obtain a prepared slide of **sperm** cells. Focus the slide first under low and then high power. These cells are quite small, so be patient and adjust your light properly.

15. Draw a picture of a sperm cell (remember to make your sketch large). Label the **cell membrane.**

16. What other structure do you notice in the sperm cell? This is called a **flagellum.** Label it on your drawing. Make an hypothesis as to the function of the flagellum.

17. What organelle, present in all the other cells, seems to be missing in both the red blood cell and the sperm cell?

DATA:

Part I:

5._____

_____

_____

10._____

_____

Part II

3._____

_____

5._____

6._____

_____

9._____

_____

12._____

_____

16._____

_____

17._____

CONCLUSIONS:

1. List the organelle(s) that were observed only in plant cells. Based on this, what function(s) can plant cells perform that animal cells can't?

2. Were there any structures found in animal cells that are not present in plant cells? What functions do these organelles confer on their cell?

3. Based on your observations, are plant or animal cells more varied in their structure? Why?

4. Based on their structure, suggest a possible function for red blood cells, sperm cells and neurons.

5. Although you did not see a nucleus in the sperm cell, it is present, but too small to be seen with your microscope. The red blood cell, however, does not have this structure. What does this information tell you about the life–span of a red blood cell?

6. Based on your observations, make a generalized statement about the structure of plant cells.

7. Do the same as above for animal cells.

8. Complete the chart below:

Organelle	Primary Function	Usually found in	
		Plants	Animals
Nucleus			
Cell Membrane			
Cytoplasm			
Cell Wall			
Central Vacuole			
Chloroplast			
Flagellum			

SUGGESTIONS FOR FURTHER STUDY:

- Investigate the structure of other plant cells, and relate any differences/similarities in organelles to the function of the cells you are examining. Prepare a slide of a paper–thin section of a potato. Observe the cells both with and without iodine stain. (Iodine turns blue/black in the presence of starch). You may also prepare slides of tomato skin and tomato pulp. Observe these cells without stain. Make drawings of all cells you see, including labels. Report on any special organelles seen, and relate these to the function of the cell.

- Examine prepared slides of other types of animal cells. Various cells that may be available include muscle (smooth, skeletal and cardiac), bone, eggs, white blood cells, and skin. Answer the same types of questions as those above.

WHAT ORGANIC NUTRIENTS ARE FOUND IN FOODS?

INTRODUCTION: Carbohydrates, lipids, proteins and **vitamins** are but some of the **organic** compounds essential to living organisms. They are found in all our cells, and must constantly be supplied, in one form or another, through the foods we eat. Carbohydrates include the substances we normally call **sugars** and **starches**. These are our primary source of energy. Proteins, composed of varying numbers of different amino acids, are the most important structural compounds in our bodies. The fatty acids found in lipids not only act as a "back–up" energy source, but also play important roles in the body's insulation, protection, and synthesis of cell membranes and steroids (including cholesterol and hormones). There are approximately 13 essential vitamins that we can not synthesize ourselves. These we must obtain from food. Vitamins, or micronutrients as they are sometimes called, are organic compounds other than lipids, carbohydrates and proteins, that the human body needs (in varying amounts) to maintain good health. One of these, ascorbic acid (vitamin C), will be investigated in this activity.

PURPOSE:

- How can you test for the presence of proteins, carbohydrates, lipids and vitamin C?

MATERIALS:

Glucose solution	Ascorbic acid
Test tubes and rack	Sudan IV dye
Corn starch solution	Boiling water bath
Iodine	Xylene, or other organic solvent
Toast	Medicine dropper
Benedict's solution	Forceps
Vegetable oil	Indophenol
Biuret reagent or concentrated nitric acid	Goggles
	Small beaker
Albumen solution	Test tube holder
Brown wrapping paper	Lab apron

PROCEDURE:

Students should wear goggles and lab apron throughout this activity.

Part I: IDENTIFICATION OF CARBOHYDRATES

1. Simple sugars, or monosaccharides, are the basic building blocks of all carbohydrates. One way to identify these in the laboratory is with a compound called Benedict's reagent. Obtain a dropper bottle of Benedict's reagent. Next, put 10 ml of glucose solution (a monosaccharide) in a clean test tube. As a control, put 10 ml of tap water into another test tube. Label the tube with the water C, for control.

2. Add a dropper of Benedict's solution to each tube, and place them in a boiling water bath for three minutes. **Wear goggles when working with boiling water.**

3. With a test tube holder, carefully remove both tubes and record their colors on Data Table 1.

4. Starches are more complex carbohydrates, composed of mono and disaccharides joined into a long chain, or polymer. Iodine is a dye which indicates the presence of starch. Get a dropper bottle of iodine.

5. Put 10 ml of corn starch into a clean test tube. Set up a control tube, as before, with 10 ml of water. Add three drops of iodine to each tube. Record the color of the two tubes on the data table.

6. Dextrins are an intermediate form of carbohydrate. They are formed when starch is partially decomposed, either through heating or chemical digestion. Obtain a piece of toast, and tear off a small section.

7. Get a small beaker of water, and add several drops of iodine. Holding the piece of toast with forceps, dip it into the beaker.

8. On your data sheet, draw a diagram showing where starch is found in the toast. What color is the iodine on the rest of the toast? Where are dextrins found in toast?

Part II: IDENTIFICATION OF PROTEIN

1. Proteins are polymers of amino acids. They differ from one another in both the numbers and types of amino acids they contain. One type of protein is called albumen (as in eggs). Obtain a test tube with 5 ml of albumen solution. Label a second test tube C, for control, and add 5 ml of water.

2. Biuret reagent is often used as a test for the presence of proteins. Get a dropper bottle of Biuret reagent. **Caution: biuret reagent is caustic, and can damage skin and clothing. Handle with care.** Add 10 drops of the reagent to both tubes. Record the color of each tube on Data Table 1.

3. Another test for proteins involves a color change when the substance being tested is combined with nitric acid. If your teacher instructs you, proceed with the next 2 steps. Otherwise, go directly to Part III.

4. Set up two test tubes, one with albumen, the other water, as you did in Step 1. Obtain a dropper bottle of nitric acid. **Caution: nitric acid is harmful to skin and clothing. Handle with extreme care.**

5. Add 10 drops of acid to each tube. Record the color of each tube on the data table.

Part III: IDENTIFICATION OF LIPIDS

1. Lipids, commonly known as fats, oils and waxes, are synthesized by combining two different types of molecules, fatty acids and alcohol. Unlike carbohydrates and proteins, they are not polymers. Pour a small amount of vegetable oil into a beaker and bring it to your lab desk.

2. One method for identifying lipids is known as the "spot test." For this test, obtain a small piece of unglazed, brown wrapping paper. Place one drop of oil on the paper, and, as a control, a drop of water in another spot.

3. After the water has dried, hold the paper up to a light and observe the two spots. Describe your results in Data Table 2.

4. Another way in which to identify lipids is to see what they dissolve in. What happens when you try to dissolve a fat in water? Test this by adding 5 ml of vegetable oil to 5 ml of water in a test tube. Shake vigorously, then let the tube stand for a few minutes. Describe the results on the data table.

5. Now add equal amounts of vegetable oil to an organic solvent, such as xylene. Shake and let settle. Describe these results.

6. Still one other test for lipids is based on their reaction with the dye Sudan IV. Into a clean test tube, place about 5 ml of water, and add 30 drops of oil. In another tube, the control, just add water.

7. Carefully add one drop of the Sudan IV dye to both tubes and shake vigorously. After the contents have settled, record the location of the dye on Data Table 2.

Part IV: IDENTIFICATION OF VITAMIN C

1. Obtain some ascorbic acid (vitamin C) in a small beaker. Next, get a dropper bottle of indophenol. Indophenol looses its color in the presence of vitamin C (the more vitamin C present, the faster the color is lost).

2. Put approximately 5 ml of indophenol into a clean test tube. Add ascorbic acid, one drop at a time. After each drop, swirl the test tube. Keep a count of the number of drops you've added. Continue until the indophenol is permanently bleached. Record the number of drops to decolorize the indophenol on Data Table 1.

Part V: IDENTIFICATION OF UNKNOWN FOODS

Ask your teacher which foods you should test. Divide each of your food samples into six equal parts. You are now going to test your sample for the presence of protein, lipid, sugar, starch, dextrins and vitamin C.

If you do not remember how to perform any of the tests required, refer back to the correct section of this investigation. If you do not remember what a positive result looks like, refer to your data tables.

For some of the tests, you will need a solution. If your sample is a solid food, you must first grind it up and dissolve it in some distilled water to make a solution. (Note: not all tests require a liquid.)

Select **two** methods of testing for lipids. Perform these on two or more samples. Record your results in Data Table 3.

If you are testing more than one food, repeat the entire process for your next unknown.

Enter all your results on the class data table. When everyone is finished, copy the class data table. You will need this information to answer some of the conclusion questions.

DATA:

DATA TABLE 1: RESULTS OF IDENTIFICATION TESTS FOR ORGANIC NUTRIENTS

Tube	Benedict's Reagent	Iodine	Biuret	Nitric Acid	Indophenol
Experimental					
Control					

Part I: 8 – Presence of Dextrins

Part III:

DATA TABLE 2: IDENTIFICATION TESTS FOR LIPIDS

Results	Spot Test	Solvent	Sudan IV
Control (water)			
Lipid (oil)			

Part V:

DATA TABLE 3: PRESENCE OF ORGANIC NUTRIENTS IN FOOD SAMPLE(S)

FoodSample	Sugar	Starch	Dextrins	Lipid	Protein	Vit.C
1						
2						
3						

CONCLUSIONS:

1. Which of the identification tests that you performed were quantitative, if any? Describe these.

2. What is a polymer?

3. Based on your class data, which was the most **complete** food tested? Why?

4. Based on your class data, which food has the most Vitamin C?

5. Based on your class data, which food(s) could be considered as "junk?" Why?

6. What's the difference between an organic and inorganic solvent? Give an example of each.

SUGGESTIONS FOR FURTHER STUDY:

- Design and perform an experiment to test the fat content of whole milk, buttermilk, skim milk, low–fat milk, heavy and light cream. Do you find any significant differences? Use a bar graph to help illustrate your results. In what way(s) might this information be valuable to you?

- Design and conduct an experiment to compare the Vitamin C content of various juices. Report your results, including a bar graph.

- Design and conduct an experiment to separate the oil from peanuts.

DIFFUSION: WHERE DO MOLECULES MOVE?

INTRODUCTION: Particles of matter are always in motion. The energy of motion is called **kinetic energy**. Thus, the amount of kinetic energy that a substance contains indicates how fast its particles move. **Gases** contain **molecules** with high kinetic energy. Their particles are constantly striking each other, rebounding, and spreading apart. In **liquids**, the molecules are also in constant motion, but with less energy than gases. If the energy is lowered even further, very little molecular movement occurs, resulting in the **state of matter** known as **solid**.

If the **concentration** of molecules of a specific type is high, they tend to collide and rebound often. The result of these collisions causes the particles to disperse as far apart as possible. This dispersion of particles is known as **diffusion**. Diffusion occurs until the particles have become concentrated equally in a given space. In this investigation, you will observe patterns of diffusion in gases, liquids and solids.

PURPOSE:

- What is the pattern of diffusion of molecules?

MATERIALS:

Test tube	Potassium permanganate crystals
Cellulose tubing	Molasses
Rubber band or string	Forceps
250–ml beaker	Goggles
Phenolphthalein	Lab apron
Ammonium hydroxide in stock bottle	

PROCEDURE:

Part I: DIFFUSION OF GASES

Wear goggles and a lab apron for this part

1. Pour approximately 10 ml of phenolphthalein into a clean test tube. Fill the rest of the tube with tap water.

2. Cut a small piece of cellulose tubing so that it will fit over the open end of your test tube. Attach the tubing with a rubber band or tie it on with a piece of string. (See illustration 1)

3. Obtain a stock bottle of ammonium hydroxide. **Use care when handling this chemical, avoid contact and do not inhale the fumes directly.**

4. Open the bottle of ammonium hydroxide, and invert the test tube over the mouth. Lower the end of the tube below the rim of the bottle. (See illustration 2)

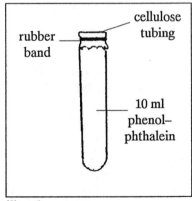

illus. 1

5. Observe the color of the phenolphthalein, and record any changes that occur.

6. **Immediately replace the top on the bottle of ammonium hydroxide when you complete your observations.**

7. Remember that phenolphthalein is a pH indicator. What can you tell about the pH of the water in the test tube? What is the pH of the ammonium hydroxide?

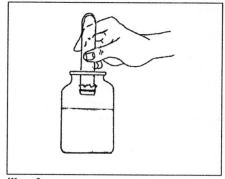

illus. 2

Part II: DIFFUSION OF LIQUIDS

1. Wash a 250 ml beaker, then carefully pour in molasses until the beaker is approximately one–half filled.

2. You are now going to add an equal amount of water, but **do not agitate the molasses** as you do so. This can be accomplished by tilting the beaker of molasses slightly, and slowly pouring the water down the side. You should have equal layers of molasses and water when you finish this step.

3. Over the next 15–20 minutes, observe the two layers carefully. Do either the water or molasses (or both) seem to diffuse?

4. Wash your hands, and dip a finger into the solution to taste the water at various time intervals. Describe your observations on the data sheet. Draw a diagram showing the original layers of molasses and water in the beaker. Using arrows, indicate the direction of diffusion of each substance.

5. Which seem to diffuse faster, gases or liquids?

Part III: DIFFUSION OF A SOLID IN A LIQUID

1. Clean your 250 ml beaker, and fill it with tap water.

2. Obtain a small amount of potassium permanganate. Using forceps, add a few crystals to the beaker of water. **Immediately** observe the results as the crystals fall to the bottom of the beaker.

3. What is the direction of diffusion of the purple color from the crystals?

4. Can you see any indication that water is diffusing into the potassium permanganate as it flows from the crystal?

5. How does the rate of diffusion of a solid compare with that of gas?

6. On the diagram on your data sheet, illustrate the water, crystals, and diffusion particles. Use arrows to show the direction of diffusion.

7. Let the beaker stand overnight. The next day, describe the distribution of the color.

DATA:

Part I: GASES

5. _____

7. _____

Part II: LIQUIDS

3. _____

4. _____

5. _____

Part III: SOLID

3. _____

4. _____

5. _____

6.

7. _____

CONCLUSIONS:

1. Explain fully why the phenolphthalein in the test tube turned red when held over the ammonium hydroxide.

2. Account for the differences in the rates of diffusion between the gas and the molasses–water mixture.

3. Define diffusion in **your own words.**

4. Where do molecules move during diffusion? What is the net result of diffusion in terms of the distribution of the particles?

SUGGESTIONS FOR FURTHER STUDY:

- The rate at which particles diffuse is affected by several variables, including **concentration of particles** and **temperature**. Using the basic experimental design that follows, test one of these variables. You will have to modify the procedure in order to include your variable. Before you begin, write your hypothesis and describe your experimental design. Have your teacher review this before proceeding.

BASIC EXPERIMENT TO MEASURE RATE OF DIFFUSION

With a single–edged razor blade, **carefully** cut several cubes from a potato. Each cube should be 1 cm on a side. Prepare or obtain a 5% potassium permangnate solution. Half–fill a small beaker with the above solution. Place several potato cubes into the beaker (see illustration 3), and record the EXACT time.

You can determine the rate at which the potassium permanganate solution diffuses into the potato cubes by removing one cube every ten minutes and slicing it in half. You will see the dye inside the cube, and can measure its penetration with a mm ruler. (See illustration 4.) The actual **rate** of diffusion can be calculated using the following equation:

$$\text{rate} = \frac{\text{distance}}{\text{time}}$$

In order to test the variable of temperature, you will need several beakers, each at a different temperature. In order to test the effect of varying concentrations, you will require several beakers with varying concentrations of potassium permanganate solution (for example, 5%, 3%, 1%).

Organize your data in a table, with the proper title and headings. Graph your results. Write a conclusion which relates to your original hypothesis.

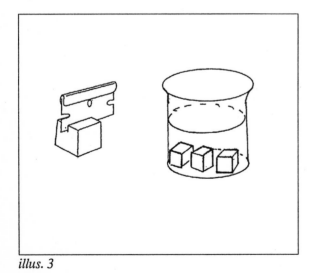

illus. 3

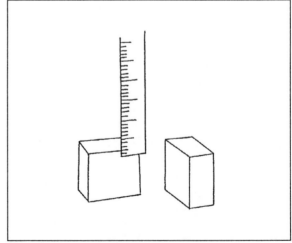

illus. 4

THE EFFECTS OF OSMOSIS

INTRODUCTION: Osmosis is a special type of **diffusion**. It occurs when water molecules diffuse across a membrane, from a region of higher concentration (of water) to a region of lower concentration (of water). Osmosis is one method by which water may enter, or leave, a living cell. Under normal conditions, osmosis is continually occurring in cells, with the amount of water leaving roughly equivalent to the amount entering. What would happen to a cell, however, if the concentration of water in its external environment decreased? This phenomonon will be illustrated using a model cell membrane, called **dialysis tubing.** The concentration of water may be decreased by adding some other substance, like salt, to the water. The more salt you add, the lower the concentration of water. You will also observe a living cell to see if it follows the same pattern as your model.

PURPOSE:

- What occurs during osmosis?

MATERIALS:

Three 250–ml beakers
3 pieces of dialysis tubing
String
10% salt solution
20% salt solution
Distilled water

2 Elodea leaves
2 glass slides
2 cover slips
Triple–beam balance
Compound microscope

PROCEDURE:

Part I: A MODEL CELL

Dialysis tubing is a synthetic membrane having tiny pores of known size. Although it is not alive, it can be used to study some aspects of osmosis as they would occur in a living cell. The tubing has been cut into useful lengths and pre–soaked so it is easy to open. If you tie a knot in one end of the tubing, you have a membrane sack which you will fill with a solution. Once filled, the other end of the tubing may be tied, and the entire "model cell" can be immersed in another solution to observe osmosis. Quantitative data can be collected by weighing the sack before and after osmosis occurs.

1. Obtain three beakers and label them 1, 2, and 3. Set them on your lab desk, you will use them shortly.

2. Obtain three pieces of dialysis tubing and securely tie-off one end of each. Using a graduated cylinder, pour 10 ml of 10% salt solution into **each** sack. Securely tie the other end of each sack as shown in the illustration on the next page.

3. Rinse each sack in tap water to remove any solution that may have spilled. Blot dry with a paper towel.

4. Select one sack, and find its mass on a triple–beam balance. This will be designated sack 1. Record its mass on Data Table 1. Place sack 1 in beaker 1.

5. Follow the same procedure for the other two sacks. After finding the mass of sack 2, place it in beaker 2. Do the same for 3. Remember to record the mass of each sack on the Data Table.

6. Into beaker 1, pour enough **distilled water** to cover the sack. The water will represent a 0% salt solution. You have just set–up a **hypotonic** situation, since the solution in the beaker contains **less solute** (dissolved salt) than the solution in the sack.

7. Into beaker 2, pour enough **10% salt solution** to cover the sack. Since the concentration of solute (salt) is the **same** in both the beaker and the sack, it is called **isotonic**.

8. Into beaker 3, pour enough **20% salt solution** to cover the sack. Since the concentration of solute is **greater** in the beaker than in the sack, this set–up represents a **hypertonic** condition.

9. What do you think the prefixes hypo–, iso– and hyper– mean?

10. Let the beakers stand for 30 minutes. During this waiting time, continue with the investigation by proceeding to Part II. **Remember to return to step 11 after 30 minutes.**

11. Remove sack 1 and blot dry. Find the mass of the sack on the same balance you used previously, and record it on the data table.

12. Repeat step 11 for sacks 2 and 3.

13. Calculate and record the percent change in mass of each sack. This can be done by following the formula below:

$$\% \text{ change} = \frac{\text{net change}}{\text{initial mass}} \times 100$$

Part II: PLASMOLYSIS IN LIVING CELLS

Normal Elodea cells are composed of approximately 99% water. The water from your sink at school is also approximately 99% "pure." In this part of the exploration you will observe what occurs when an Elodea cell is placed in an 80% water (20% salt) solution.

1. Prepare a wet mount slide of an Elodea leaf.

2. Focus the slide under the low power magnification of your compound microscope. Isolate a single cell near the edge of the leaf, and switch to high power. Adjust your light as needed.

3. On a separate sheet of unlined paper, sketch one elodea cell, paying particular attention to the location of the chloroplasts in relation to the cell wall. Label the cell wall and chloroplasts. Remove your slide, but keep it handy at your lab desk.

4. Now prepare a wet mount of another Elodea leaf, but this time, instead of water, use 20% salt solution to prepare your slide.

5. Focus on a single cell first under low, then high power magnification.

6. After waiting a few minutes, carefully observe the Elodea cell. Again, concentrate on the location of the chloroplasts in relation to the cell wall. On the same paper you used previously, sketch the cell you are now observing. Include the same labels as before. If you are having difficulty remembering the appearance of the first slide you prepared, look at it again before sketching this one.

7. What is the difference in appearance between the cell prepared with water and the one prepared in salt solution?

8. What happens to a cell when placed in a hypertonic solution?

When a cell shrinks in size due to outward osmosis it is said to be plasmolyzed. **Plasmolysis** is the term that describes this event.

DATA: Part I:

DATA TABLE 1: CHANGE IN MASS OF DIALYSIS SACKS BEFORE AND AFTER OSMOSIS

Beaker #	Solution in Sack	Solution in Beaker	Mass of Sack At Time 0 (g)	Mass of Sack At 30 Minutes (g)	Net Change in Sack Mass (g)	% Change in Mass
1	10% Salt	Distilled H$_2$O				
2	10% Salt	10% Salt				
3	10% Salt	20% Salt				

9. _____

_____

Part II:

7. _____

8. _____

CONCLUSIONS:

1. In Part I, which sack(s) increased in mass? Why?

2. Which sack(s) decreased in mass? Why?

3. What do you think was happening, if anything, in the isotonic set–up in Part I?

4. Summarize the movement of water, in terms of concentration, during osmosis.

5. Describe the changes that occur inside an Elodea cell that has undergone plas-molysis. How can you account for these changes?

6. How did the behavior of the living Elodea cell compare with your model? Which beaker did the cell's behavior most closely approximate? What is the value of using the model instead of the actual cell?

SUGGESTIONS FOR FURTHER STUDY:

- In winter, roads are sometimes salted to help melt ice and snow. In terms of osmosis, what do you think this does to the vegetation along the roadside? Why?

- People will sometimes kill slugs or worms by pouring salt on them. Why would this practice kill the organism?

- Design and conduct one or more experiments to answer any of the following questions:

    a. How does temperature affect the rate of osmosis?

    b. How does the difference in concentrations on opposite sides of a membrane affect the rate of osmosis?

    c. How does the surface area of the membrane affect the rate of osmosis?

All three of these questions may be answered using an experiment similar to the one you performed in Part I of this activity. Before you begin, write out your hypothesis and experimental design. Get approval from your teacher. Design a data table for your results. Prepare a graph of your results, with the variable being tested (temperature, for example) along the horizontal axis, and the net change in mass on the vertical axis.

PAPER CHROMATOGRAPHY

INTRODUCTION: Have you ever wondered how scientists figure out what chemicals are found in various solutions? One method that is used to separate different compounds in a mixture is called **chromatography**. In this technique, a complex mixture is dissolved in an appropriate **solvent**. The solvent is then allowed to **diffuse** through an absorbent medium, such as filter paper. Since the different components of the mixture will vary in their chemical and physical properties, each will *settle out* of solution in various places. When the solvent has traveled over the entire surface of the paper, each individual component will be isolated at a specific location along the paper strip. Once isolated, the separate compounds can be identified by other chemical or physical tests.

In this exploration, you will separate the various compounds found in **chlorophyll**. You will find that although chlorophyll, the **pigment** essential for photosynthesis, appears to be one homogeneous, green substance, it is actually a mixture of several different colored compounds.

PURPOSE:

- To separate the component pigments in chlorophyll through the process of paper chromatography.

MATERIALS:

Spinach leaves
Hot plate
1000 ml beaker
250 ml beaker
Alcohol
Chromatography paper
Large test tube & cork
Test tube rack
Hair blower (optional)

Chromatography solvent
Scotch tape or thumb tack
Beaker tongs
Forceps
Capillary tube
Metric ruler
Funnel
Goggles

PROCEDURE:

Before you can separate the compounds of chlorophyll, you must first prepare a concentrated chlorophyll solution. Chlorophyll can be removed from leaves by boiling them in hot alcohol. When following the directions below, remember the following. **Alcohol is flammable: Never heat alcohol on an open flame. Never put alcohol directly on a hot plate. To heat alcohol, use a "double boiler;" a small beaker containing the alcohol, placed inside a larger beaker containing water. Place the larger beaker on the hot plate. Do not allow the alcohol to boil over the rim of the beaker. Wear goggles when conducting this activity.**

1. Place 3–4 large spinach leaves in a small beaker. Add only enough alcohol to cover the leaves. Place this beaker inside a 1000 ml beaker that has been half–filled with water. Place the two beakers on a hot plate to boil the water. (See illustration 1)

2. When all the chlorophyll has been bleached from the leaves, carefully remove them from the beaker using forceps. Discard the leaves. Continue boiling the alcohol until the color of the chlorophyll is a **very dark green.** Do not be concerned if a lot of the alcohol evaporates, since you only need a small amount of extract for the activity. Using tongs, remove the beakers from the hot plate and allow the chlorophyll extract to cool.

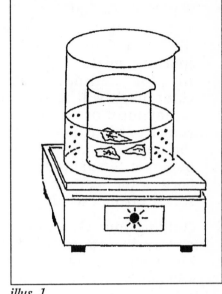

illus. 1

3. While the extract is cooling, prepare your chromatography apparatus. Cut a piece of chromatography (filter) paper so that it's approximately the same length as a large test tube. Obtain a cork that fits your test tube, and tape, or tack, one end of the paper to the bottom of the cork, as show in illustration 2. Put the cork and paper into the test tube. Be sure the paper **just touches the bottom of the tube.** Cut the paper if it is too long. If the paper is more than a few millimeters short, cut another piece to fit properly.

4. Remove the paper from the tube and place it on a clean, flat surface. With a sharp pencil, draw a horizontal line 2 cm from the bottom of the paper. You are now ready to apply your chlorophyll extract.

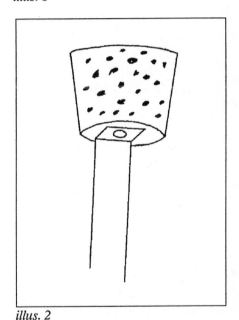

illus. 2

5. Using a capillary tube, place one drop of extract in the center of your pencil line. Do not allow the drop to spread more than a few millimeters. Allow the spot to dry (blow on it or wave the paper in the air). Add another drop of extract on top of the first. Again, allow the spot to dry. Repeat this procedure several times, until you have a dark, concentrated spot of chlorophyll extract. **The more concentrated your spot of chlorophyll is, the more accurate will be your results.**

6. Allow your chlorophyll spot to dry completely before proceeding. If a hair blower is available, this can be used to quickly dry the extract.

7. Place your test tube in a test tube rack. Using a funnel, carefully pour one cm of the solvent mixture into the bottom of the tube. **Take care not to inhale the solvent mixture directly. Keep the mixture tightly stoppered whenever possible.** Do not let the solvent drip down the sides of the tube. Insert your cork–paper apparatus into the tube, so that the end of the paper is *in* the solvent, but the spot of extract is *not.* (See illustration 3). Be sure the cork is securely inserted into the tube.

8. Watch as the solvent makes it way up the paper. The leading edge of the diffusing solvent is called the "front." As the front passes your chlorophyll spot, the mixture dissolves in the solvent. As absorption continues, various components of the chlorophyll will settle out of the solution at different locations along the paper. When the solvent front is approximately 1 cm from the top of the paper, remove the paper from the test tube. Separate the paper from the cork, and replace the cork in the test tube to prevent the vapors of the solvent from escaping.

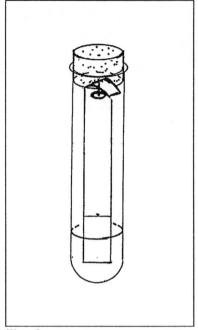

illus. 3

9. Holding your **chromatogram** by the edges, wave it gently in the air to dry. Once dry, place the chromatogram on a flat surface. Quickly mark the solvent front with a pencil line. Also draw a pencil line across the center of each colored spot which appeared. This is necessary since the colors tend to fade. Tape your chromatogram to the data sheet. Write the color of each of the separated compounds on the data sheet, directly adjacent to its place on the chromatogram.

One way to tentatively identify the chemicals isolated on a chromatogram is by their R_f values. R_f stands for "rate of flow," which is a measure of how fast the compound traveled in relation to the speed of the solvent. This figure can be calculated for each pigment according to the following equation:

$$R_f = \frac{\text{distance pigment travelled}}{\text{distance solvent travelled}}$$

Calculate the R_f values for each of the pigments you isolated. Measure the distance of each pigment, to the nearest .1 cm, from the lower pencil line to the line drawn at the center of each pigment. Measure the distance travelled by the solvent by using the lower and uppermost pencil lines. Record your calculations in Data Table 1.

The names of the plant pigments you isolated on your chromatogram are **carotene, xanthrophyll (a & b), chlorophyll a** and **chlorophyll b.** Experimental evidence shows that the R_f values decrease from carotene to chlorophyll b, as listed above. Write the names of your pigments, as determined from their relative R_f values, on the data table.

DATA:

Chromatogram

[ ]

DATA TABLE 1: R$_F$ VALUES OF CHLOROPHYLL PIGMENTS

Color of Pigment	Distance Solvent Traveled (cm)	Distance Pig- ment Traveled (cm)	R$_f$ Value	Name of Pigment

CONCLUSIONS:

1. According to your data, how many different pigments are found in chlorophyll? What are their colors?

2 . Why were each of the pigments deposited at different locations on the chromatogram?

3. Could you identify chlorophyll a on the chromatogram of another student performing the same experiment? Explain.

4. Using the information you gained in this investigation, explain what might happen to leaves in the autumn that would cause them to turn various shades of red, orange and yellow.

5. What is a solvent?

SUGGESTIONS FOR FURTHER STUDY:

- Construct chromatograms, using the same procedures as you did in this investigation, of leaves from various plants. If possible, use leaves of varying colors or shades of green. For each chromatogram prepared, identify the leaf from which it was made, and the initial color of the leaf. Using R_f values, identify the pigments isolated on each new chromatogram. Do all leaves contain the same pigments? Report your findings.

- Differing proportions of chemicals in the solvent mixture produce varying results. The solvent you used in this investigation contained 92 parts petroleum ether and 8 parts acetone. Prepare other solvent mixtures by varying the amount of acetone (between 5 and 30 parts). Another chemical that could be combined with the petroleum ether to prepare a solvent mixture is 5–15 parts ethyl or isopropyl alcohol. Still other solvents may be prepared by combining 85–95 parts xylene or toluene with 5–15 parts ethyl or isopropyl alcohol, or acetone.

  Prepare a chromatogram in the same manner as you did in this investigation, but use your own solvent mixture. Better yet, try 2 or 3 different solvents. Use the same types of leaves in each of your new chromatograms. Compare your results. Determine the R_f values of each pigment for each different chromatogram. How does the nature of the solvent affect the R_f values of each pigment?

- If you live near a seacoast, obtain samples of green, red or brown algae. Analyze the pigments present in the algae by preparing a chromatogram. Use a reference guide to help identify your sample. What conclusions can you draw about the variety of pigments found in algae? Report your conclusions, and include your chromatograms.

MEASUREMENT AND ALTERATION OF pH

INTRODUCTION: When certain substances dissolve in water, they tend to break apart (**dissociate**) and release **ions**, atomic particles with either a positive or negative charge. Every substance which dissociates does not release the same number of ions. An **acid** is a substance which releases positively charged **hydrogen ions** in solution (H^+). The more ions released, the stronger the acid. A **base** is a substance which releases negatively charged **hydroxyl ions (OH⁻)**. The more hydroxyl ions released, the stronger the base.

The concentration of ions present in a particular solution is measured in terms of **pH**. The pH scale runs from 0 to 14. The <u>lower</u> the reading on the pH scale, the more H^+ are present in the solution, and the stronger the acid. Conversely, the <u>higher</u> the number on the scale, the more OH^- ions are present, and the stronger <u>is the base</u>. If the amount of H^+ and OH^- are equal, the pH reading is in the middle, or 7. A solution with a pH of 7 is said to be **neutral**. A solution with a pH between 0 and 6 is **acidic**, while one with a pH between 8 and 14 is considered **basic**.

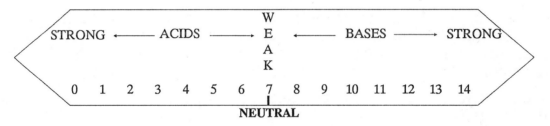

Most organic reactions require a specific pH range. Therefore, most living organisms, both plant and animal, require a specific internal or external pH for their survival. Since you will be studying many biological reactions this year, it will often be important to know how acidic or basic a particular solution is. To determine pH, specific dyes, called **indicators** are used. Indicators change color at a specific pH. Some indicators are so sensitive that they detect only a narrow range of pH, while others merely detect the presence or absence of ions.

Usually, if acid is added to a substance, it's pH will decrease (lower numbers indicating more acidity). And if a base is added, the pH usually rises. A **buffer** solution, however, resists changes in pH, even with the addition of acid or base. Since living systems must maintain a narrow pH range in order to survive, they use buffers. This ability to prevent large changes in pH is an extremely important characteristic of living organisms.

PURPOSE:

- In what ways can pH be measured, altered, and buffered?

MATERIALS:

Distilled water
.1 N hydrochloric acid (HCl)
Graduated cylinder
.1 N sodium hydroxide (NaOH)
Small beakers
Test tubes & rack
Phenolphthalein
Milk

Assorted biological materials, such as
  milk, fruits or juice, egg, vegetables
  or juice, vinegar, soil samples
Litmus paper
Bromthymol blue
pH (hydrion) paper
Goggles
Lab apron

PROCEDURE:

Part I: INDICATORS

1. **Wear goggles and a lab apron.** Clean three test tubes and place them in a rack at your lab desk. Label test tube 1 acid, test tube 2 neutral, and test tube 3 base.

2. Into test tube 1, pour 10 ml of the hydrochloric acid solution (HCl). Into test tube 2, pour 10 ml of distilled water (H_2O). Into test tube 3, pour 10 ml of the sodium hydroxide solution (NaOH).

3. Get a dropping bottle of phenolphthalein. This is one of the indicators you may be using during the year. Add several drops of the indicator to each of your three test tubes. (Add the same number of drops to each tube). Record the color of the indicator as it appears in each tube on Data Table 1.

4. Wash out your test tubes. Refill them with the same solutions as you used in the first step.

5. Get a dropping bottle of bromthymol blue indicator. Repeat the procedure, entering your observations in the Data Table. Clean out your test tubes.

   The next two indicators you are going to test have been saturated onto papers, so it will be easier to test these by dropping the acid, base, and neutral solutions **onto** the papers.

6. Obtain dropping bottles of HCl, NaOH, and H_2O. Get three pieces of red and three pieces of blue litmus paper, and place them on a paper towel at your lab desk.

7. Add one drop of each solution to a different piece of red and blue litmus paper. Record the color of the paper (and hence the indicator) on the data table. Be sure to dispose of the used litmus papers properly.

8. Obtain some pH paper with its color chart. Your pH paper may come either in rolls or individual strips. If rolled, tear off three pieces of about 3 cm. each (do not waste this paper!) Place the papers on a paper towel in front of you.

9. Add a drop of each solution to the papers in the same manner as before. Match the color of the paper to the color chart, and record the pH **number** on your data table.

Part II: MEASURING THE pH OF BIOLOGICAL SUBSTANCES

1. Select the indicator that you found to be the most specific. Using this indicator, measure the pH of the various biological materials available. Record your data in Data Table 2. *Remember, a substance must first be in solution in order to determine its pH. If you are trying to test a solid, you must first dissolve some in distilled water, then test the pH of the solution.*

Part III: ALTERING pH AND INVESTIGATING A BUFFER

1. Put 25 ml of tap water in a small beaker. Using pH paper, record its pH in Data Table 3.

2. Get a dropper bottle of HCl. Add one drop to the tap water in the beaker. Carefully swirl the beaker to mix. Measure the pH and record the data. Continue to add one drop of HCl, mix, and measure and record pH, until you have added a total of 10 drops of acid.

3. Continue this process, but measure the pH of the water after every five drops, until you have added a total of 30 drops of acid to the water.

4. Wash the beaker thoroughly, and refill with another 25 ml of tap water.

5. Repeat the procedure using the base, NaOH. In Data Table 3, record the pH after each drop for the first ten drops, then after every 5 drops until 30.

6. You will now perform the same activity, but instead of tap water, you will use a biological solution. Fresh liver has been mixed with distilled water in a blender. (This is known as liver homogenate). Pour 25 ml of this homogenate into a clean beaker. Record the pH on Data Table 4.

7. First add HCl, one drop at a time, and record the pH of the liver homogenate after each drop. Continue, as before, until you have added 10 drops. Continue adding drops of acid, until 30, but record your data after every 5 drops added.

8. Wash your beaker thoroughly, and refill with 25 ml of fresh homogenate.

9. Repeat the entire procedure adding drops of NaOH. Record the data on Table 4.

10. Plot the data from Tables 3 and 4 on the same axes of a graph. Label the horizontal axis "drops of acid/base." Label the vertical axis "pH." Use a pen to plot the points for the tap water, and a pencil to plot the data for liver homogenate. Use a solid line for the acid curve, and a broken line to illustrate the data for the base. Give your graph a title.

DATA:

DATA TABLE 1: REACTION OF INDICATORS TO ACID, BASE AND NEUTRAL SOLUTIONS

Solution	Phenolphthalein	Bromthymol Blue	Litmus	pH paper
ACID (HCl)				
NEUTRAL (Water)				
BASE (NaOH)				

DATA TABLE 2: pH of BIOLOGICAL MATERIALS

Substance Tested →					
pH →					

DATA TABLE 3: CHANGES In pH of TAP WATER

Number Drops HCl or NaOH	pH of	
	Acid + H_2O	Base + H_2O
0		
1		
2		
3		
4		
5		
6		
7		
8		
9		
10		
15		
20		
25		
30		
Total pH Change		

DATA TABLE 4: CHANGES IN PH OF LIVER HOMOGENATE

Number Drops HCl or NaOH	pH of	
	Acid + Liver	Base + Liver
0		
1		
2		
3		
4		
5		
6		
7		
8		
9		
10		
15		
20		
25		
30		
Total pH Change		

CONCLUSIONS:

1. List the indicators you used to determine pH. Which is the most specific? Why?

2. Which of the substances tested in this investigation are acids? List them in order of decreasing strength.

3. Which of the substances tested are bases? List them in order of decreasing strength.

4. Is it possible to make a conclusion regarding the pH of biological materials? If so, what is it? If not, why not?

5. Review the graph you prepared in Part III. Make a summary statement regarding this data.

6. Which substance used in this investigation is a buffer? How can you identify it as such?

7. In order to maintain proper health, your blood's pH must remain within a very narrow range (7.0–7.8). Why don't you get sick every time you drink some orange juice?

SUGGESTIONS FOR FURTHER STUDY

- If your school owns a pH meter, use it to measure the pH of the same substances you tested in this investigation. How does your data compare?

- If you live near any bodies of water, test a sample and determine its pH. Research the topic of acid rain, and determine if this condition is affecting your geographical area.

- The mathematical equation for determining pH of a solution is:

$$pH = -\log [H^+] = \frac{\log 1}{[H+]}$$

Using a chemistry/math book to help you, explain the pH scale in terms of this equation. What is meant by pOH?

WHAT DO HYDROLYTIC ENZYMES DO?

INTRODUCTION: Enzymes are specialized protein compounds that are essential to living systems. These **organic catalysts** regulate the speed of biochemical reactions. Enzymes function in all chemical reactions, including cellular respiration, synthesis of needed compounds, and chemical digestion. Since chemical digestion involves the **hydrolysis** of larger molecules into smaller ones, the enzymes involved are known as hydrolytic enzymes. If complex food molecules were not reduced to a simpler form, they would be unable to pass through the membranes of the digestive tract and enter the bloodstream.

Hydrolytic enzymes, like all others, are specific for the reactions they catalyze. Some will act only on one class of compounds. Others will recognize only one particular **substrate**. Still others will catalyze only one specific chemical reaction. In this investigation, you will determine the function of two specific hydrolytic enzymes, **pepsin** and **salivary amylase**.

PURPOSE:

- What is the function of hydrolytic enzymes?

MATERIALS:

Hard boiled egg	1 saltine cracker
Distilled water	5 ml of diastase or saliva
Single–edged razor blade	Lugol's iodine solution
Pepsin solution	Benedict's solution
Dropper bottle of HCl	Boiling water bath
Test tubes, Test tube rack	Goggles
pH paper	Triple–beam balance
Dropper bottle of NaOH	Incubator or oven

PROCEDURE:

PART I: PROTEIN HYDROLYSIS

Egg white is almost pure protein. In this section you will investigate the action of pepsin, a **protease**, on pieces of egg white.

1. Obtain a hard boiled egg. Peel the shell, and separate the white from the yellow part of the egg. Discard the yellow.

2. With a single–edged razor blade or scalpel, carefully cut the egg white into four small cubes, approximately 5 mm on a side. Try to make the cubes as close to the same size as possible.

3. Place four clean test tubes in a rack, and number them 1–4.

4. Using a triple–beam balance (or, if available, an analytical balance), find the mass of one of the egg white cubes to the nearest tenth of a milligram. Record its mass as cube 1 in Data Table 1. Place the cube into test tube 1.

5. Repeat step 4 for each of the other three cubes. Remember to record the mass, then place the cube in the appropriate test tube.

6. Into each of the four test tubes, add the following:

> Tube 1 .... 10 ml of distilled water
> Tube 2 .... 10 ml of pepsin solution
> Tube 3 .... 10 ml of pepsin + 10 drops of HCl
> Tube 4 .... 10 ml of pepsin + 10 drops of NaOH

7. Using pH paper, determine the pH of the contents of each of the four test tubes. If the pH of tube 3 is not below 4, add more HCl untilit is. If the pH of tube 4 is not above 9, add more NaOH until it is. Record the pH of each tube on the data table.

8. Label the test tube rack with your initials, and place it in an incubator or oven at body temperature, approximately 37°C, 48 hours.

PROCEED TO PART II OF THE INVESTIGATION: RETURN TO STEP 9 AFTER 48 HOURS.

9. Remove your test tube rack and examine each cube for complete or partial hydrolysis. On the Data Table, record the appearance of each of the tubes and egg white cubes.

10. Carefully remove each cube (one at a time) and gently blot dry on paper towels. Find the mass of the cube, using the same balance you used in step 4. Record the new mass on your data table.

11. Empty and clean your test tubes.

PART II: STARCH HYDROLYSIS

Starch is a complex carbohydrate. Its presence may be detected by a blue–black color change in Lugol's iodine solution. Sugars are simple carbohydrates, and may be identified by their reaction with heated Benedict's solution. In this section you will investigate the action of **salivary amylase** (found in saliva) or **diastase** (a plant amylase) on starch.

1. Get one saltine cracker, two test tubes, and dropper bottles of iodine solution and Benedict's solution.

2. Break the cracker into quarters. Test one quarter for the presence of sugar by crumbling it into a test tube and adding 10 ml of distilled water and 2 ml of Benedict's solution. Shake to mix. Place the tube in a boiling water bath for 3 minutes. **Always wear goggles when boiling liquids.** Record your results in Data Table 2. Is sugar present in the cracker?

3. Test another quarter for starch by placing one or two drops of Lugol's iodine solution directly on the cracker. Record its color on Data Table 2. Is starch present in the cracker?

4. Crumble the two remaining quarters into two separate test tubes. Label one of the tubes C, for control, and add 10 ml of distilled water. To the other tube, add 5 ml of diastase or saliva, and 5 ml of distilled water. Shake each tube to mix.

5. Label the tubes with your initials, and place them in an incubator or oven at 37°C for 30 minutes.

6. Now test the contents of both tubes for starch and sugar. Do this by removing

a small amount of material from each tube and testing it for starch. Then remove another small amount from each tube and test this for sugar. Remember that the Benedict's test for sugar requires heating for 3 minutes in a boiling water bath. Record your results on Data Table 2.

DATA:

DATA TABLE 1: RESULTS OF PEPSIN ON EGG WHITE (PROTEIN) CUBES

Cube #	pH	Mass (g) Start	Mass (g) Final	Appearance
1				
2				
3				
4				

DATA TABLE 2: RESULTS OF AMYLASE ON STARCH (CRACKER)

Nutrient	Preliminary Cracker	Control Tube	Experimental Tube
Sugar			
Starch			

CONCLUSIONS:

1. What is the substrate on which pepsin acts?

2. What is the substrate on which amylase acts?

3. What substance(s) would you expect to find in the test tubes that showed partial or complete hydrolysis of the egg white? Why?

4. According to your data, how does pH affect pepsin action?

5. What is the function of a protease, such as pepsin?

6. In Part I of the activity, which tube(s) served as a control? How?

7. How are sugars and starches related, chemically?

8. What is the function of an amylase, such as saliva or diastase?

9. When a sick person is given intravenous feeding in the hospital, glucose solution is usually the food used. Why would the doctor recommend glucose instead of starch?

SUGGESTIONS FOR FURTHER STUDY:

-   Do other organisms contain hydrolytic enzymes to digest food? Design and conduct an experiment to see if yeast contain enzymes to hydrolyze sucrose, a dissacharide, into glucose, a monosaccharide. Follow the guidelines below in planning your investigation. When you have written an hypothesis and an experimental design, have it approved by your teacher before proceeding. Organize your data and analyze your results.

    An active yeast suspension may be prepared by dissolving a package of active, dry yeast in 1 liter of warm water, several hours before it is needed.

    Sucrose is found in ordinary table sugar.

    Benedict's solution shows a color change when boiled for 3 minutes with glucose, but no color change with sucrose.

    Boiling kills yeast cells.

    Benedict's solution can show approximate glucose concentrations in the various colors it changes:

light green	less than .25% glucose
yellow–green to yellow	.25% – 1.0%
orange	1.0% – 2.0%
brick red	over 2.0%

WHAT VARIABLES AFFECT ENZYME ACTION?

INTRODUCTION: Enzymes are specialized **protein** molecules that **catalyze** biochemical reactions. One particular enzyme, called **amylase**, hydrolyses large starch molecules into smaller **monosaccharide** molecules. This enzyme is essential to the chemical digestion of complex **carbohydrates**. The rate at which this, and other enzymes function, however, is dependent upon various factors. In this exploration you will investigate four variables that affect the rate of amylase action on starch. These factors are **temperature, pH, substrate concentration,** and **enzyme concentration.**

PURPOSE:

- What are the effects of changes in temperature, pH, enzyme concentration and substrate concentration on the activity of amylase?

MATERIALS:

7 test tubes	Lugol's iodine solution
95% ethanol	Medicine droppers
Test tube rack	Buffer solutions in dropper
Distilled water	bottles: pH 3, 8, 11
Thermometer	Test tube holder
.05% amylase solution	Goggles
Ice	Graph paper
1% starch solution	Incubator/oven
Hot water baths	Colorimeter (optional)

PROCEDURE:

Part I: EFFECT OF TEMPERATURE

1. Place 4 clean test tubes in a rack, and number them 1–4. To each tube add 5 ml of distilled water and 20 drops of starch solution. Shake each tube to mix the contents.

2. Pre–incubate each of the tubes at the following temperatures for **20 minutes:**

    1: 0°C   **2:** 38°C   **3:** 70°C   **4:** 100°C

A beaker of ice water can serve as a temperature bath for tube 1. Tubes 2 and 3 must be placed in separate beakers and warmed to the proper temperature on a hot plate. A boiling water bath is appropriate for tube 4. **Always wear goggles when boiling water.**

3. While the tubes remain in their water or ice baths, add 2 ml of amylase solution to each tube. Keep the tubes in their temperature baths for an additional 10 minutes.

4. At the appropriate time, remove the tubes from their temperature baths and return them to the test tube rack. **Use a test tube holder when removing hot tubes.** Stop any chemical reaction that is occurring in the tubes by adding a

dropper–full of ethanol to each tube. Mix well. (Alcohol will stop the action of the enzyme).

5. Allow the tubes to return to room temperature. While you are waiting, prepare two control tubes. To one tube add 5 ml of distilled water, 20 drops of starch solution, and 2 drops of Lugol's iodine solution. Label this tube C–1. To the other control tube, add 5 ml of distilled water and 2 drops of iodine. Label this tube C–2.

6. Once the tubes have returned to room temperature, add 2 drops of iodine solution to each. Shake well.

   Lugol's iodine tests for the presence of starch. The more starch, the darker the iodine solution (blue–black shows starch present). A yellowish color indicates the absence of starch.

7. What are the colors of the two control tubes? What does this indicate?

   In order to determine the relative rate of enzyme action in each of the 4 experimental tubes, you will compare their colors with the two control tubes. On a color scale from 1 through 5, Control tube–1 is to be rated a "5" (darkest color; most starch present), and Control tube–2 is to be rated as "1" (lightest color; least starch present).

8. Assign a numerical rating from 1 to 5 to each of your experimental tubes. Record this data on Data Table 1.

9. Which number on your rating scale shows the MOST enzyme action? Why? Which number indicates the LEAST amount of enzyme action? Why?

10. On a separate sheet of graph paper, construct a graph showing the effects of temperature on enzyme action. On the horizontal axis, plot a scale of temperature from 0°C–100°C. On the vertical axis, show "Enzyme Action" using your rating scale of 1–5. Remember to label each axis and title your graph.

Part II: EFFECT OF pH

1. Wash and dry your experimental test tubes. Keep your control tubes. You will use them again for this part.

2. Set–up three experimental test tubes as follows:

   Add 5 ml of the following buffer solutions:

   1: pH 3   2: pH 8   3: pH 11

   To each tube add 20 drops of starch.

3. Which tube is the most acidic? The most basic?

4. Add 2 ml of amylase solution to each tube. Swirl the tubes to mix the contents. Put your initials on the test tube rack, and place the rack with the test tubes in an incubator or oven, set at body temperature (approximately 37°C), for 20 minutes.

While you are waiting, proceed to Part III. Return to step 5 after 20 minutes.

5. Remove the rack from the incubator.

6. Add two drops of iodine to each tube and mix well. Compare their colors to the control tubes. Assign each tube a rating from 1–5 as you did in Part I. Record this data in Data Table 2.

7. Prepare a graph of your results from this section. Title the graph "Effect of pH on Enzyme Action." Label the horizontal axis "pH," and the vertical axis "Relative Rate of Enzyme Action."

Part III: EFFECT OF ENZYME CONCENTRATION

1. Clean and dry four test tubes. Keep the control tubes to use again in this section. Prepare the four tubes with 5 ml of distilled water and 20 drops of starch solution in each.

2. Place the test tubes in a rack, and add the following amounts of enzyme (amylase solution) to each tube:

    1: 1 drop   2: 10 drops   3: 40 drops   4: 60 drops

3. Mix the contents of each tube, and place the rack with the test tubes in the incubator at body temperature for 20 minutes.

Proceed to Part IV while the tubes are incubating. Return to step 4 after 20 minutes.

4. Remove the tubes from the incubator. Add 2 drops of iodine solution to each tube and mix well.

5. Again, assign color ratings to each tube, using your control tubes as a reference. Record your data on Data Table 3.

6. Prepare a graph of your results, entitled "Effect of Enzyme Concentration on Enzyme Activity." Along the horizontal axis, plot enzyme concentration in drops. Label the vertical axis as you did for Parts I and II.

Part IV: EFFECT OF SUBSTRATE CONCENTRATION

1. Obtain five clean test tubes, keeping the controls to use again. Repeat the experimental procedure as described in Part III, but vary the amount of starch solution, instead of the amount of amylase used, as follows:

    **1: 5 drops starch   2: 20 drops starch   3: 50 drops starch
    4: 75 drops starch   5: 100 drops starch**

2   What is meant by the term substrate? What is the substrate for amylase?

3   In the appropriate data section, describe the contents of tubes 1–5 as you prepare to put them in the incubator.

4. Follow the same procedure as in Part III. Record your data in Data Table 4.

5. Prepare a graph of these results. Include the proper title and labels.

DATA:

Part I:

7._____

DATA TABLE 1: EFFECTS OF TEMPERATURE ON AMYLASE ACTIVITY

Temperature °C	0	38	70	100
Color Rating Scale 1 – 5				

9._____

_____

Part II

4._____

DATA TABLE 2: EFFECTS OF pH ON AMYLASE ACTIVITY

pH	3	8	11
Color Rating Scale 1 – 5			

Part III:

DATA TABLE 3: EFFECTS OF ENZYME CONCENTRATION ON AMYLASE ACTIVITY

Enzyme Concentration drops	1	10	40	60
Color Rating Scale 1 – 5				

PART IV:

2._____

3._____

_____

_____

DATA TABLE 4: EFFECTS OF SUBSTRATE CONCENTRATION ON AMYLASE ACTIVITY

Substrate Concentration drops	5	20	50	75	100
Color Rating Scale 1–5					

CONCLUSIONS:

1. What is the effect of temperature on the rate of enzyme activity?

2. What is the optimum temperature for amylase?

3. What is the effect of changes in pH on the rate of enzyme activity?

4. What is the optimum pH for amylase?

5. The pH of stomach secretions is quite low, due to the presence of hydrochloric acid. What does this information tell you about starch digestion in the stomach?

6. Describe the effects on enzyme activity of increasing either substrate or enzyme concentrations.

SUGGESTIONS FOR FURTHER STUDY:

- Design and conduct an experiment to more accurately determine the optimum temperature or pH for starch hydrolysis. Record and graph your results.

- Design and conduct an experiment to determine the optimum pH for the hydrolysis of proteins by the enzyme pepsin. NOTE: Protein solutions may be prepared with albumen. Biuret reagent turns lavender in the presence of protein.

- Using reference materials, such as biochemistry books or scientific articles, explain the effects of high temperatures and/or alcohol on the structure and action of proteins (enzymes).

THE EFFECT OF BILE ON FAT

INTRODUCTION: Lipids, commonly known as fats and oils, are not **soluble** in water. Therefore, lipids are not easily digested by the **lipase** in the watery environment of your digestive tract. Lipid molecules, in addition, tend to form small, solid spheres, with very little **surface area** exposed for enzyme action. How then does your body digest butter, cream or oil? In this activity, you will investigate the action of **bile**, a **secretion** that is produced by your **liver**, stored in your **gall bladder**, and functions in your **small intestine.**

PURPOSE:

- What is the effect of bile on the digestion of fats?

MATERIALS:

5% bile solution
4 test tubes
2% lipase solution
Test tube rack
Olive oil

Water bath (37°C)
Distilled water
NaOH solution
Phenol red solution

PROCEDURE:

1. Place four clean test tubes in a rack, and number them 1–4.

2. Add the following materials to each of the test tubes:

   #1

   10 ml of distilled water
   2 drops of oil

   #2

   5 ml of distilled water
   5 ml of lipase solution
   2 drops of oil

   #3

   5 ml of distilled water
   5 ml of bile solution
   2 drops of oil

   #4

   5 ml of distilled water
   5 ml of bile solution
   5 ml of lipase solution
   2 drops of oil

3. In your data section, describe the appearance of each tube.

4. Write an hypothesis about what you think will happen in each of the tubes.

5. In order to determine if digestion occurs, you will use the indicator phenol red. Place 20 drops of the phenol red in each of the test tubes. Shake well. If the color of any of the tubes is not pink, add the base sodium hydroxide, one drop at a time, until the color remains pink.

6. What is the color of phenol red in the presence of a base?

7. Place all of the tubes in a water bath at body temperature, 36°– 40°C.

8. Observe your tubes after 10, 20, and 30 minutes. Write your observations in Data Table 1.

9. Phenol red turns yellow in the presence of acid. In which of your tubes is acid present? In which did it appear earliest?

DATA:

Tube	Appearance at Start of Experiment
1	
2	
3	
4	

4._____

_____

_____

_____

6._____

DATA TABLE 1: RESULTS OF INCUBATION OF OIL WITH VARIOUS SOLUTIONS

Tube	Contents of Tube	Results at 10 Minutes	Results at 20 Minutes	Results at 30 Minutes
1	OIL WATER			
2	OIL WATER LIPASE			
3	OIL WATER BILE			
4	OIL WATER BILE LIPASE			

9._____

CONCLUSIONS:

1. How do you account for the acid condition found in the tubes that turned yellow? (HINT: What are the end products of lipid digestion?)

2. What other substance would you expect to find in the tubes(s) that contain acid?

3. Is there any evidence that bile alone is a digestive enzyme? Explain.

4. According to your results, in what way does bile affect lipid digestion?

5. The actual process by which bile causes the effects you observed is called **emulsification.** Using a reference book if needed, define emulsification.

6. People occasionally require surgery to remove their gall bladder. How does this affect their ability to digest lipids?

SUGGESTIONS FOR FURTHER STUDY:

- Using current scientific references, research the relationship between dietary intake of fats and oils, and cholesterol build–up in the blood. Why is the level of cholesterol in your blood important? Report on your findings in a brief paper (including bibliography).

- Design and conduct an experiment to compare the rate of digestion of plant vs. animal lipids. Common plant oils include corn, "vegetable," sunflower and soy bean. Bacon, butter, eggs, and cream are some examples of foods containing animal fat. Report your results.

- Using references such as biochemistry or organic chemistry text books, describe the differences between saturated and unsaturated fats.

- Using scientific references, distinguish between high–density lipoproteins (HDL's) and low–density lipoproteins (LDL's). What is their significance?

DOES YOUR DIET MEET YOUR NUTRITIONAL NEEDS?

INTRODUCTION: How healthy is your diet? Should you take vitamin supplements? How many calories do you consume in a day? Is it true that you should reduce your intake of fats? These are just some of the questions that people have been raising in recent years, as scientists have shown that the relationship between diet and health is more than coincidental. Dietary factors are implicated in many diseases, including cancer. Obesity has been shown to decrease one's lifespan significantly. Vitamin deficiency diseases are still common in many parts of the world. But how do you know what a "healthy" diet should contain? How much protein is actually needed by your body on a daily basis? How much vitamin A? The Food and Nutrition Board of the National Academy of Sciences, in Washington, D.C., has published a list of **recommended dietary allowances (RDAs)**. These figures show the amounts of protein, vitamins and minerals required for the maintenance of health in individuals of all ages. (See Chart 1 in Data section).

In this investigation, you will record all the food and drink you consume in one or more days. You will then calculate the amounts of various **nutrients** in these foods, determine your totals for each day, and compare your diet with the RDAs listed for your age and sex.

PURPOSE:

- Does your diet meet your nutritional needs?

MATERIALS:

One or more diet worksheets
Reference sources for nutritional value of foods

PROCEDURE:

1. Your teacher will tell you how many days you are to include in your diet record. Obtain a diet worksheet for each day you will be collecting data.

2. Beginning as soon as you wake up on Day 1, list **all** foods and drinks (other than water) you consume during the day. Be sure to indicate how much of each food you eat in the column labeled "Amount." The units you use for recording the amount of any food will vary depending upon the food. A carrot, for example, may be listed as "1," while a portion of french fried potatoes may be recorded as "1 cup," or "1 average portion." If in doubt as to the amount, just describe it in the most logical terms available. **Remember to include everything you eat and drink, from the time you wake up to the time you go to sleep!**

3. If you eat foods from a container, such as a box, can or wrapped package, look for any nutritional data that may be supplied. Very often nutrient amounts will be listed on the package, and the quantity of calories, protein, carbohydrates, etc. will be given. If this information is available on your container, copy it down before throwing away the package. This is the type of data you will need later, in order to analyze your diet.

4. When you have completed gathering your data, use the references available to record the nutritional information requested on the worksheet. You may already have some of this data from the package labels you copied. If not, just look up each item of food or drink, and adjust the quantities of each nutrient listed, according to the amount of that food you actually consumed. For example, if the data for bread is given per slice, and you had two slices on a sandwich, be sure to multiply the figures given by 2.

5. Calculate your totals, for each day, for each data column given. If your data include more than one day's food intake, find the **daily average** for each column of data.

6. Compare your totals in calories, protein (grams), vitamin A (I.U.), thiamin (mg), riboflavin (mg), niacin (mg) ascorbic acid (vitamin C) (mg), calcium (mg), phosphorus (mg) and iron (mg), to the RDAs listed on chart 1. Analyze your diet, in terms of recommended nutritional value, in the data section.

   NOTE: You should be aware that the figures given in chart 1 are valid for an **average sized** boy or girl. Specifically, the figures reflect the RDAs for a male of 145 lbs. (66 kg) and 69 in. (176 cm) tall, and a female weighing 120 lbs. (55 kg), and 64 in. (163 cm) tall. If your height, weight or age varies *significantly* from this "standard," you will have to adjust the RDAs slightly. See your teacher for help with these adjustments.

DATA:

DATA CHART 1: SELECTED RDAs

SELECTED ITEMS FROM *RECOMMENDED DIETARY ALLOWANCES* (WASHINGTON,D.C.: NATIONAL ACADAMY OF SCIENCES, 1980)											
SEX	AGE	CALORIES	PRO-TEIN (g)	VIT. A (I.U.)	THI-AMIN (mg)	RIBO-FLA-VIN (mg)	NI-ACIN (mg)	VIT. C (mg)	Ca (mg)	P (mg)	Fe (Iron) (mg)
Males	15 - 18	2500 - 3500	56	1000	1.4	1.7	18	60	1200	1200	18
Females	15 - 18	1500 - 2500	46	800	1.1	1.3	14	60	1200	1200	18

Analysis of your diet:

CONCLUSIONS:

1. In what respects is your diet healthy?

2. In what way(s) is your diet unhealthy?

3. What specific changes could you make in your diet to correct any imbalances you observed?

4. Using reference material, describe the 4 (or 5, if you have a very recent source) food groups. How may these groupings help you in planning a healthful diet?

5. Carbohydrates and fats are both nutrients that are essential to the proper functioning of your body. Why don't you think there are any RDAs for these compounds?

6. A general standard for carbohydrate and fat intake lists 300 g and 70 g, respectively, for girls, and 400 g and 90 g for boys. How do your totals for carbohydrates and fats compare with these figures? How, specifically, could you lower or raise each of these totals if needed?

7. Define the nutritional significance of each of the following substances:

   a) vitamin:

   b) mineral:

   c) carbohydrate:

   d) protein:

   e) fats:

   f) calories:

SUGGESTIONS FOR FURTHER STUDY:

- There are people who do not believe that the government's RDAs are sufficient. They feel that ever larger quantities, especially in the case of various vitamins and trace minerals, would enhance an individual's health. Investigate the claims of those who profess "megavitamin" therapy. Report on your findings. What is your opinion regarding megadoses in a daily diet and their value for specific illnesses?

- Investigate several "fad diets" that have gained popularity in recent years. Who are the people proposing these diets? What is their background? Are the diets nutritionally "safe?" Report on your findings.

- Research the problem of world hunger. How does disease and death relate to poor nutrition, especially in developing, or Third World countries? What is meant by the terms "famine," "malnutrition," "undernutrition," and "starvation?" Report your information.

DIET WORKSHEET

NAME _____   DATE _____

Food Or Drink	Amount	Calories	Protien (g)	VitaminA (IU)	Thiamin (mg)	Riboflavin (mg)	Niacin (mg)	Ascorbic Acid (mg)	Calcium (mg)	Phosphorus (mg)	Iron (mg)
Daily Total											

YEAST FERMENTATION: THE CLASSROOM WINERY

INTRODUCTION: Wine, in fact all alcoholic beverages, are produced by the action of microscopic organisms, such as yeast and bacteria. If you feed some grape mash to these organisms, they will **ferment** the sugar in the grapes, obtaining energy for themselves (in the form of **ATP**), and releasing waste products into their environment. One of these wastes, the organic compound **ethyl alcohol (ethanol)**, provides the beer, liquor and wine industries with their revenue! Different types of alcoholic beverages may be prepared by "feeding" the yeast or bacteria different types of mash. Another waste produced during **anaerobic respiration** is the gas carbon dioxide (CO_2). The baking industry makes use of this waste product to cause bread to rise. In this investigation, you will observe the anaerobic fermentation process of yeast, and produce your own grape or molasses "wine."

PURPOSE:

- What are the products of yeast fermentation?

MATERIALS:

Yeast suspension	2 bent glass tubes
2 thermos bottles	Two 2–hole stoppers to fit
Molasses or grape juice	thermos bottles
2 thermometers	Two 1–hole stoppers to fit flask
Pyrogallic acid	Glycerin (to insert glass and
2 small test tubes	thermometer into stoppers)
Flask with limewater	Graph paper

PROCEDURE:

1. Set up a fermentation apparatus as shown in the illustration on the next page. Half fill the thermos bottle with equal parts of yeast suspension and molasses (or grape juice). Attach a string to a small test tube so that it will dangle in the thermos. Fill the small tube with pyrogallic acid, a chemical that will absorb the oxygen from the air in the thermos bottle.

    The limewater is an indicator that will become cloudy if CO_2 is produced.

    Record the initial temperature reading in Data Table 1.

2. What is the purpose of removing oxygen from the thermos bottle?

3. How will you determine if energy is being produced in the thermos? (HINT: heat is one form of energy).

4. As a control, set up another apparatus, identical to the first, but **without** the yeast. Label this C, for control. Record the initial temperature of the control thermos.

5. Observe both your experimental and control set–ups over the next 48 hours. During these observations, record the temperatures of both bottles, and describe any changes seen in the limewater. The more observation you can make during the 48 hour period, the more accurate will be your data.

6. After 48 hours, remove the stopper from each thermos. Smell the contents. What do they smell like? Can the odor of alcohol be detected in either bottle?

7. Using your data, construct a graph showing the temperature changes in both the experimental and control bottles over 48 hours. On the horizontal axis plot "time in hours," and on the vertical axis plot "temperature in $^{\circ}$C." Use different colored lines for the experimental and control data.

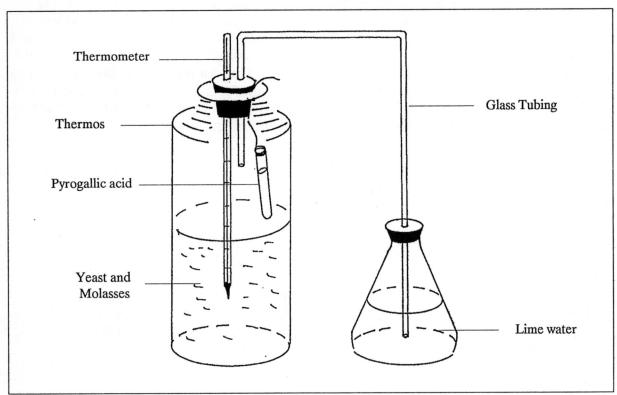

DATA:

2._____

3._____

6._____

DATA TABLE 1: RESULTS OF YEAST FERMENTATION

Time (Hours)	Experimental Bottle		Control Bottle	
	Temperature °C	Limewater	Temperature °C	Limewater
0				

CONCLUSIONS:

1. What evidence is there that a reaction is occurring in the experimental thermos? Do your observations of the control bottle verify this?

2. What evidence is there that the reaction occurring in the thermos is fermentation?

3. What is the food source for the yeast?

4. According to your data, what are the end–products of fermentation?

5. Interpret the graph you prepared. Why did the temperature rise in the experimental bottle? Give two reasons that could account for the temperature drop at the end of the time period.

6. Write a simplified equation for the process of fermentation.

SUGGESTIONS FOR FURTHER STUDY:

- Design and conduct an experiment to determine the best food source for yeast to ferment. Any mixture providing a carbohydrate source, such as corn, potato, apple, or orange mash may be tried. Report your results.

- Design and conduct an experiment to determine the best temperature for yeast fermentation. Obtain teacher approval for your idea before beginning your experiment. Report your results.

- Research the specific processes used in the wine industry. How do these procedures differ for red and white wines? Prepare a paper presenting your findings.

BAKE–A–BREAD: VARIABLES OF FERMENTATION

INTRODUCTION: During the process of **anaerobic respiration (fermentation)**, **yeast** metabolize nutrients in order to produce energy, in the form of **ATP**. As a result of this process, waste products are also produced. The production of one of these wastes, **carbon dioxide**, is the basis of the baking industry. In order to make bread rise, yeast are allowed to "eat" some of the ingredients. As they do this, they produce energy for themselves, and release carbon dioxide gas into their environment (the dough). It is the escape of this gas which causes the bread to rise. In this activity you are going to investigate a few **variables** of yeast fermentation by baking bread.

PURPOSE:

- What variables have a significant affect on the process of yeast fermentation in bread baking?

MATERIALS:

Triple–beam balance
Graduate cylinder
Materials required for baking bread: see recipe at end
Graph paper

PROCEDURE:

Your teacher will help you get organized into groups of four. Each group will investigate one variable that may affect the fermentation process of the yeast, and therefore alter the results of your baked product. A list of possible variables to investigate includes:

amount of yeast
amount of sugar
time allowed for rising
amount of kneading
temperature during rising

As a group, decide which variable you will test, and register this with your teacher. Write an **hypothesis**, on your data sheet, as to how this variable may affect yeast fermentation, or your finished product.

At the end of this lab, you will find a recipe for white bread. The recipe states the "proper" conditions for each variable listed above. One member of each group will follow these instructions. This will serve as the **control** for your experiment. Other members of the group will each alter the "proper" amounts as stated in the recipe. For example, suppose that your group is testing the amount of sugar used. If the recipe calls for 1 cup of sugar, that is what should be used by the student conducting the control experiment. Student 2 should cut the stated amount of sugar in half (1/2 cup). Student 3 should double the stated amount of sugar (2 cups). And student 4 should add NO sugar at all. From the results of your finished breads, you should be able to determine if the amount of sugar does, in fact, affect the process of fermentation in yeast.

On Data Table 1, indicate how each member of the group is altering the stated recipe.

Once you have selected your variable, written an hypothesis, and assigned experiments to each member of the group, obtain permission from your teacher to proceed with your bread baking. Your teacher may instruct you to carry out this part of the investigation at home, or in the home economics room. Wherever it is done, however, take care to follow the directions precisely. You want to ensure that any results you achieve are actually due to the variable you altered, and not to sloppy or careless techniques. Before you begin, read through the recipe. If there are any terms with which you are unfamiliar, look these up in a cookbook, or ask the home economics teacher for help. Be sure you have all the supplies and equipment needed to follow the recipe.

In order to determine whether your variable did, in fact, alter the results of the baking process, it will be necessary to compile some quantitative data. In this case, you will measure the mass, volume and density of your bread before and after baking. If you are doing the actual baking of the bread at home, the "before baking" measurements must be coordinated with other members of your group, so that you all are working on the same time schedule. If you are using an oven at school, try to take the "before baking" measurements immediately before you put your bread in the oven. Write this data in Table 2 on your data sheet. Collect the following data:

Mass: Weigh a plastic bag, to the nearest .01 g.
          Put your ready–to–bake dough in the plastic bag.
          Weigh the bag with the dough, again to the nearest .01 g.

Volume: Fill a container, large enough to hold your dough, with a measured
          amount of water.
          Immerse your plastic bag, containing the ready–to–bake dough, in
          the water.
          Measure the volume of water displaced.

Density: Calculate according to the following formula:

$$\textbf{Density} = \frac{\textbf{mass}}{\textbf{volume}}$$

After you have baked your bread, repeat these measurements (using the same plastic bag and same balance, if possible) and enter the data in Table 2.

Each member of the group should prepare a bar graph showing mass, volume and density measurements for the "pre baking" and "post baking" data they collected. Use the information on the graphs and data tables from all members of the group to analyze your hypothesis. Use the class data and graphs to help you answer the conclusion questions that follow.

DATA:

Variable being tested by your group _____

Hypothesis: _____

DATA TABLE 1: EXPERIMENTAL CONDITIONS FOR EACH GROUP MEMBER

NAME				
Experimental Condition				

DATA TABLE 2: MASS, VOLUME AND DENSITY OF "PRE" AND "POST" BAKING BREAD

Data	Pre Baking	Post Baking
Mass (g)		
Volume (ml)		
Density (g/ml)		

CONCLUSIONS:

1. Did you support or negate your hypothesis? What is your conclusion?

2. What role does yeast play in the baking process?

3. What role does sugar play (aside from taste) in the baking process?

4. How does temperature affect the process of fermentation?

5. How do you think the author of the recipe determined the best quantity of yeast to use?

6. Alcohol is another waste product of yeast fermentation. Why don't you get "drunk" everytime you eat bread?

RECIPE: PERFECT WHITE BREAD

1/2 package active dry yeast   1/8 cup warm water
1 cup milk, scalded   1 tablespoon sugar
1 teaspoon salt   1/2 tablespoon shortening
3 cups sifted, all–purpose flour

Soften active dry yeast in warm water (40° C) for about 5 minutes. Combine hot milk, sugar, salt and shortening. Cool to lukewarm. Stir in 1 cup of the flour; beat well. Add the softened yeast; mix. Add enough of the remaining flour to make a moderately stiff dough (on very humid days, you may have to add even more flour than is called for in the recipe. Turn out on slightly floured surface; knead till smooth and satiny (8–10 minutes). Shape in a ball; place in lightly greased bowl, turning once to grease the surface. Cover; let rise in a warm place until double in size (about 1 hour). Punch down. Shape into a smooth ball; cover and let rest 10 minutes. Shape into a loaf and place in a greased loaf pan. Cover and let rise until double in size (30 to 45 minutes). ** Bake in a hot oven (400°F) 35 minutes, or until done.

Be aware of safety when baking. Remove hot objects with care.

** At this point, take your "pre baking" measurements.

SUGGESTIONS FOR FURTHER STUDY:

- Conduct an experiment to determine whether sugar or starch solutions support yeast fermentation more completely.

Prepare six fermentation tubes, like the one in illustration 1. Into each tube place equal amounts of 10% solutions of glucose, lactose, maltose, sucrose, corn starch and rice starch.

Label each tube. Add 0.5 ml of bromthymol blue to each tube. Insert cotton plugs in the tubes, and sterilize in an autoclave for 15 minutes at 15 lb.

illus. 1 Fermentation tube

Allow the tubes to cool. Using a sterile pipette, inoculate each tube with 2 ml. of yeast suspension. Replace the cotton plugs, and place the tubes in a warm place.

Observe the tubes for the next 2 days. If fermentation has occurred, the tubes will look like the one in illustration 2. (Why?)

illus. 2

Record the color of the bromthymol blue, and the amount of gas produced, in millimeters, for each tube. Draw a bar graph showing the millimeters of gas produced in each tube. Remove the cotton plugs and smell the solutions. Record any odors. Write up your experiment. What conclusions can you draw? What is the function of bromthymol blue?

WHAT HAPPENS DURING BREATHING?

INTRODUCTION: The basic function of the **respiratory system** of any animal, including you, is to supply the body's cells with oxygen and remove the waste gas, carbon dioxide. These two gases are exchanged as a result of **aerobic cellular respiration,** the process by which nutrients are oxidized for energy. In humans and many other animals, gases are brought into and out of the body via the **lungs,** by the mechanical process of **breathing.** This process depends upon changes in air pressure. If the atmospheric pressure is greater than that of the lungs, air will flow in and fill the lungs. This event is called **inspiration.** When the pressure in the lungs is greater than that of the atmosphere, **expiration** occurs, and air, including the waste gas carbon dioxide, flows out of the lungs.

In order to change the pressure within the lungs, the size of the **thoracic cavity** (chest) is altered by the action of specific muscles. (See figures A and B below). The muscles of the rib cage (**intercostal**) work with the muscle that makes up the "floor" of the thoracic cavity, the **diaphragm,** to control the size of the cavity, and thereby alter the pressure in the lungs. During inspiration, the diaphragm contracts, lowering the floor of the thoracic cavity. At the same time, the intercostal muscles also contract, raising the rib cage, and further increasing the size of the cavity. When the muscles are allowed to relax and return to their normal resting state, the size of the thoracic cavity decreases, allowing expiration to occur. Although you do have some conscious control over the contraction of these muscles, they are usually directed by nerves under the control of the **medulla.** This is the respiratory center of your brain, and automatically adjusts your **breathing rate** to maintain homeostasis when carbon dioxide or oxygen levels are altered.

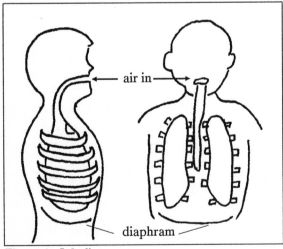

Figure A: Inhaling

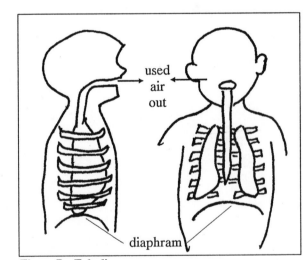

Figure B: Exhaling

PURPOSE:

- What are the major events that occur during breathing?

MATERIALS:

Tape measure (metric if available)   Limewater
Stethoscope   3 test tubes with rack
70% alcohol solution   Straws or 10 ml pipettes
Cotton

PROCEDURE:

Part I: CHANGES IN SIZE OF THE THORACIC AND ABDOMINAL CAVITIES DURING BREATHING

1. Place the tape measure around your chest at a level just below your arm pits. Measure the size of your thorax immediately after each of the following breaths:

    a. normal inspiration .... inhale normally
    b. forced inspiration .... inhale as deeply as possible
    c. normal expiration.................................. breathe out normally
    d. forced expiration.................................. "blow out" all the air you
                                                      possibly can

Record your results on Data Table 1.

2. Place the tape measure around your abdomen, at a level just below the "bottom" of the sternum, near your "last" (floating) ribs. Measure the size of your abdomen immediately after each of the different types of breaths, as you did for step 1. Record your measurements in Data Table 1.

3. Answer the following questions on the data sheet:

    a. When did the thorax and the abdomen have the greatest measurements?
    b. When did the thorax and abdomen have the smallest measurements?

Part II: SOUNDS DURING BREATHING

The flow of air is often silent, but sometimes a sound is produced as air rushes through narrow passageways or reaches a dead end. **Bronchial breathing sounds** are often heard when air rushes through the **trachea** and bronchi. **Vesicular breathing sounds** can be heard when air fills up the **alveoli** in the lungs. Listening for respiratory sounds is a diagnostic tool to detect and pinpoint respiratory disorders.

1. Clean the earplugs of a stethoscope with the alcohol solution before and after using.

2. Place the cup of the stethoscope just below your larynx and listen for bronchial sounds as you take deep breaths. Record any sounds you hear on Data Table 2.

3. Place the cup of the stethoscope on your chest in the areas listed below and listen for rustling, vesicular sounds while taking deep breaths.

    a. various intercostal (between the ribs) spaces
    b. below the clavicle
    c. below the scapula (shoulder blade)

Record any sounds you hear on Data Table 2.

4. Answer the following questions on your data sheet:

    a. Which sounds were heard during both inspiration and expiration?
    b. Which sounds were heard only during inspiration?

Part III: GAS EXCHANGE DURING BREATHING

Receptors in your blood constantly monitor levels of oxygen and carbon dioxide. Increasing levels of CO_2 stimulate the respiratory center of the brain and cause an increase in breathing rate. With an increased breathing rate, more CO_2 can be released. Conversely, decreasing levels of O_2 also cause an increased breathing rate. Therefore, cellular respiration (metabolism) determines the rate of **external respiration,** or breathing.

Limewater, or calcium hydroxide, will be used as an indicator to measure the amount of carbon dioxide in exhaled air. CO_2 forms a white precipitate in limewater, turning it cloudy. The more CO_2 present in your breath, the faster the limewater will turn cloudy.

1. Label three test tubes 1 – 3. Fill each about 1/3 full of limewater.

2. Using a straw or pipette, bubble your exhaled breath into the solution in test tube 1 until it turns cloudy. Measure the time it takes for the limewater to become cloudy, and record this in Data Table 3. **Exhale gently.**

3. Breathe deeply and quickly for about 2 minutes. **Stop if you become dizzy.** Immediately after this rapid breathing, bubble your exhaled breath into the limewater in test tube 2. Measure the time it takes for the limewater to turn cloudy, and record your results on Data Table 3. Allow your breathing to return to normal before continuing.

4. If you are in good health, exercise vigorously for about 5 minutes. Check with your teacher as to the type of exercise you should do. Immediately after the exercise, repeat the bubbling procedure with test tube 3. Record the time it takes to turn the limewater cloudy.

5. Answer the following questions on your data sheet:

    a. Which test tube showed the greatest quantity of CO_2 exhaled?
    b. Which test tube showed the least?

DATA:

DATA TABLE 1: SIZE OF THORAX AND ABDOMEN DURING BREATHING

| Body area | BREATHING MEASUREMENT | | | |
| | Inspiration | | Expiration | |
	Normal	Forced	Normal	Forced
Thorax				
Abdomen				

3a. _____

b. _____

DATA TABLE 2. BREATHING SOUNDS

| Breathing sounds | AUDIBLE SOUNDS DURING | |
	Inspiration	Expiration
Bronchial		
Vesicular		

4a. _____

b. _____

DATA TABLE 3. SUMMARY OF EXHALED CO_2

Test tube	Activity	Time To Turn cloudy	Relative Amount of CO_2 *
1	Normal Breathing		
2	Hyperventalation		
3	Excercise		

* highest, lowest, middle

5a. _____

b. _____

CONCLUSIONS:

1   Describe the muscle actions involved in normal inspiration and exhalation.

2.   Which muscles, other than those mentioned in your answer to question 1, seem to be involved in forced exhalation? On what evidence is your answer based?

3.   Which breathing sounds, recorded in Data Table 2, might change in a patient who has pneumonia? Bronchitis? Why?

4.   Explain why each of the activities listed in Data Table 3 resulted in the relative amount of CO_2 recorded.

5.   Describe how the level of CO_2 in the blood regulates the breathing rate.

SUGGESTIONS FOR FURTHER STUDY:

- Prepare a model of the human thoracic cavity that will demonstrate the mechanics of breathing. This model may be constructed according to illustration 2. Describe how your model reflects the structures involved in external respiration. Demonstrate your model to the class. In what way is your model inaccurate?

- Research one of the following diseases that affect the respiratory system: pneumonia, emphysema, asthma, bronchitis, lung cancer, cystic fibrosis. Prepare a report that details the effects on the respiratory system caused by the condition on which you are reporting. Include a current bibliography with your report.

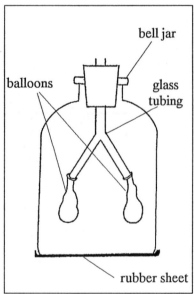

illus. 1

THE HEMACYTOMETER SIMULATOR

INTRODUCTION: Hematology is the study of blood, and deals primarily with the number and types of red and white cells. Red blood cells, known as **erythrocytes**, are responsible for transporting oxygen throughout the body. White blood cells, or **leukocytes**, come in many different shapes and sizes, but all are involved with the process of immunity, or the protection of the body from foreign antigens.

Normally, there are 4,500,000–6,000,000 erythrocytes per mm^3 (one drop) of blood in the adult human. Leukocytes are far less numerous, with only 5,000–10,000 per drop of blood. Although the normal range for the numbers of these cells is rather broad, a **blood cell count** is an important diagnostic tool when illness is suspected, or general health is to be checked.

Technicians perform blood cell counts using a specialized microscope slide called a **hemacytometer** (See illustration 1). Etched into this slide are two, 9 mm square areas called **counting chambers**, that are sub–divided into tiny, lined squares of known size. (See illustration 2). The smallest squares are a mere .0025 mm! These divisions make counting cells a much easier job than it would be without them. Both red and white blood cells can be counted in these chambers. Of course, a technician can not possibly count millions of cells, so a small sample is counted. The actual blood count is then determined by multiplying the sample count by a number that takes into account the amount of dilution as well as the volume of the counting chamber.

Since it is impractical to perform an actual blood count in the classroom, you will simulate this activity in this exploration.

counting chambers

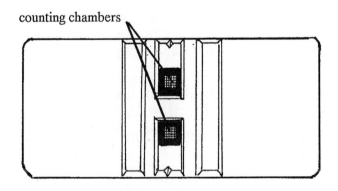

illus. 1: Hemacytometer

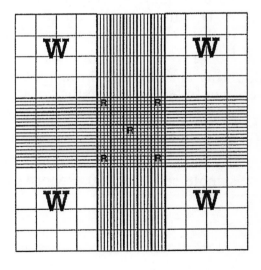

illus. 2: Counting chambers

PURPOSE:

- To interpret data from red and white blood cell counts.

MATERIALS:

Biology reference book

PROCEDURE:

1. Look at the Sample Counting Chamber in the data section. This represents what you might see if you looked through a microscope, on high power, at a hemacytometer chamber. If you were performing a red cell count, this square would represent one of the small rectangles labeled "R" in illustration 2.

2. Count the cells in each small box of this counting chamber. Follow a pattern when counting, such as the one suggested in illustration 3. Count only those cells lying completely within a box, and those touching the upper and right lines of each box. Do not count those touching the left hand and lower lines. In the Sample Counting Chamber, for example, you should count 12 red blood cells.

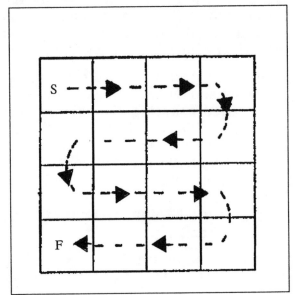

3. Count the number of red blood cells represented in the erythrocyte counting chambers A through D on the data sheet. Record your results in Data Table 1.

illus. 3: Counting pattern

4. To determine the actual number of red blood cells in a mm^3 of each sample A–D, multiply the number of cells counted in each chamber by 100,000 (1.00x10^5). Record the actual red blood cell count for each sample in the data table.

Counting chambers E through H represent white blood cell samples. These chambers would be located in the corners labelled "W" in illustration 2.

5. Count the number of leukocytes represented in samples E through H on the data sheet. Record these results in Data Table 2.

6. To determine the actual number of white blood cells/mm^3, multiply your sample count by 100. Record these calculations in the data table.

DATA:

SAMPLE COUNTING CHAMBER A

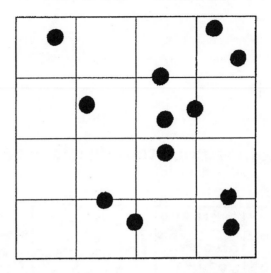

ERYTHROCYTE COUNTING CHAMBERS A–D

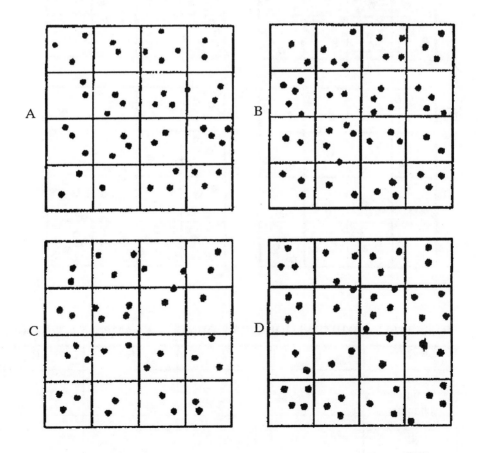

DATA TABLE 1: RED BLOOD CELL COUNTS FOR SAMPLES A THROUGH D

# Cells	A	B	C	D
In Counting chamber				
In mm^3 Sample				

LEUKOCYTE COUNTING CHAMBERS E – H

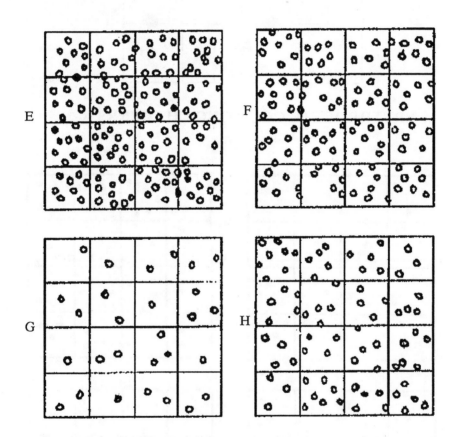

DATA TABLE 2: WHITE BLOOD CELL COUNTS FOR SAMPLES E THROUGH H

# Cells	E	F	G	H
In Counting chamber				
In mm^3 Sample				

CONCLUSIONS:

1. Using a biology reference text, draw a typical red and white blood cell.

2. Which of the counting chambers represent samples with blood counts in the normal range? (Indicate whether these are erythrocyte or leukocyte counts).

3. Using a reference text, describe the condition known as anemia. Which, if any, of the counting chambers might represent a sample from an anemic individual?

4. When you get an infection your body responds by producing more white blood cells than normal. This temporary elevation in the white blood cell count helps doctors diagnosis these infections. Which sample(s) could have come from an infected individual?

5. An individual undergoing radiation therapy (x–ray treatment for conditions such as cancer) often has a lower than normal white count. Which sample could represent such an individual?

6. Using a biology reference text, determine the general hematological symptoms of leukemia. Which sample might be from an individual with leukemia?

SUGGESTIONS FOR FURTHER STUDY:

- Research the various forms of anemia or leukemia. Describe their physiological symptoms as they relate to hematology. Report on your results.

- A differential white blood cell count is a specialized method used to determine the numbers of various types of white blood cells. Describe the structure and function of at least 5 different leukocytes. Using a prepared slide, perform a differential w.b.c. count with at least 100 w.b.c's in your sample (you may have to use more than one slide.) Compare your results with the standard for each w.b.c. given below. What conditions might be present in individuals with abnormal levels of these various leukocytes? Report your results.

  neutrophils 55–70%
  eosinophils .... 1–3 %
  monocytes 3–8%
  lymphocytes 20–30%
  basophils 0.5–1%

WHAT FACTORS AFFECT CAPILLARY CIRCULATION?

INTRODUCTION: Your blood, and that of many other animals, constantly flows through your body from **arteries**, to the smaller **arterioles**, to **capillaries**, back through **venuoles** and finally into **veins**. Your heart is the pump that initiates each surge of bloodflow. Your brain and autonomic nervous system are in charge of regulating the rate at which the blood flows through your body. The rate of blood flow is most easily altered by changing the diameter of the blood vessels or changing the rate at which your heart beats. What factors might cause your blood flow to speed up or slow down? In this investigation, you will determine the effects of various chemicals and temperatures on the rate of circulation in the capillaries of a goldfish. The goldfish is used for this activity because the "skin" in its tail is so thin that the blood vessels below can easily be seen under a compound microscope.

PURPOSE:

- What are some factors that affect capillary circulation?

MATERIALS:

Goldfish
Petri dish
2 "half slides"
Compound microscope
Ice chips

Chemical solutions (see your teacher)
Warm water (40°C)
Small beaker of fish tank water
Medicine dropper

PROCEDURE:

1. Prepare a strip of wet cotton that you will wrap around your goldfish. As long as the cotton remains wet, and covers the goldfish's gills, the fish can remain out of water for quite a long time. Use the water from the goldfish tank to soak your cotton. The illustration shows you how your fish will look when it is wrapped in the cotton. (Do not cover the fish's mouth with the cotton).

2. Bring your cotton and petri dish to the fish tank. With a net, remove a healthy goldfish from the tank. Remove the fish from the net and wrap the cotton strip around the fish's gills. Gently place the fish in the petri dish. Do not be concerned if it flops around a bit. Just be sure to adjust the cotton wrapping if it becomes dislodged. Have a small beaker of tank water and a medicine dropper handy. You will have to add drops of water to the cotton in order to keep it wet.

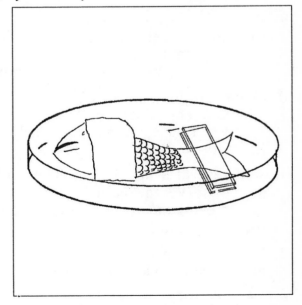

3. Place one of the half–slides under the goldfish's tail. Put the other half–slide on top. If the fish moves its tail, you may have to readjust the slides.

4. Remove the stage clips from a compound microscope. Place the petri dish on the stage, so that the tail of the fish is directly under the objective lens. Using low power magnification, move the petri dish around until you can see blood vessels. Focus on a vessel with the high power magnification.

5. Observe the pattern of blood flowing within the vessels. You should see blood moving in two general directions. The arterioles carry blood away from the heart, and the flow in these vessels is relatively rapid.Venuoles carry blood back toward the heart, and the flow is much slower. The capillaries are the smallest vessels, and they branch from the arterioles and connect back to the venuoles. Look for these capillaries. On the data sheet, make a sketch of each type of vessel. Use arrows to indicate the direction of blood flow.

6. Look closely at the **red blood cells.** These are the small blood cells traveling through the vessels. Sketch a red blood cell on the data sheet.

7. About how many red blood cells can pass through a capillary at one time? How does the diameter of the red blood cell compare to the diameter of the capillary?

8. Get a small amount of crushed ice. Add a few pieces to the goldfish's tail, under the slide. Observe the capillaries through the microscope for several minutes.

   Look for any change in capillary diameter and speed of blood flow. A good way to determine if the capillary diameter has changed is to observe the size of the red blood cells within the vessel. Since they do not change size, you can determine if the vessel has **dilated** or **constricted** by comparing how much room the cells take up now, as compared to the way they fit before you added the ice. Write your observations on the Data Table. Flood the fish's tail with aquarium water so that the circulation returns to normal.

Remember to keep the cotton surrounding the goldfish wet!!

9. Add a few drops of warm water (40°C) to the fish's tail. Again, observe the capillary diameter and blood flow for a few minutes. Record your observations on the Data Table. Add several drops of regular aquarium water to the fish's tail to get the blood flow back to normal.

10. Get a dropper bottle of nicotine or alcohol (or another substance of your teacher's choosing). First, be sure your goldfish seems in good shape. If the blood is flowing extremely slowly, or the cotton has dried–up, replace your fish with another from the tank.

11. Add several drops of the chemical to the fish's tail. Wait about one minute, and observe the capillary circulation. How did the chemical affect the diameter of the vessels and the rate of the blood flow? Write your results in the Data Table.

12. Flood the fish's tail with aquarium water so that the circulation returns to normal. **Be sure to keep the cotton wet.**

13. Obtain a dropper bottle of adrenaline or acetylcholine (or another chemical supplied by your teacher). Repeat the procedure you just followed using the new chemical. Record your results.

14. Carefully return your fish to the tank.

DATA:

Sketches of venuole, arteriole and capillaries:

Sketch of red blood cell:

7. _____

DATA TABLE: EFFECTS OF VARIOUS FACTORS ON CAPILLARY CIRCULATION:

Substance added	Capillary Diameter	* Rate of Blood Flow	* More or Less Than Normal
ice			
warm water			

CONCLUSIONS:

1. What is the effect of cold temperatures on capillary blood flow?

2. Which chemical(s) made capillary changes like those described in question 1?

3. What do you think the effect on your body would be if you ingested the chemical listed in question 2 while you were out in the cold?

4. People often look "flushed," or red when they are hot, or overheated. According to the results of your experiment with warm water, how can you explain this flushing of the skin?

5. Both adrenaline and acetylcholine are natural body chemicals that are secreted when your body needs more blood going to various organs, and needs to get the blood there faster. How do your results confirm the role of these substances?

6. You should have observed that red blood cells pass through a capillary almost one at a time. What is the advantage of having blood cells pass through a capillary in "single file?" (Hint: what is the function of these cells?)

7. How do the diameters of arterioles, venuoles and capillaries compare? Based on these observations, where do you think the majority of transfer of materials between cells and blood occurs?

SUGGESTIONS FOR FURTHER STUDY:

- Extend this investigation to include other chemical substances. Make a list of chemicals that you suspect may cause vasodilation or vasoconstriction. Show this list to your teacher and have it approved. Conduct your experiment and report your results.

- Investigate research conducted on the effects of nicotine (smoking) or alcohol (drinking) on the circulatory system. Report on your readings.

- Design and conduct an experiment to determine the effects of various substances or conditions on pulse rate. The pulse is a reflection of the rate at which your heart beats. What conditions or substances would increase this rate? What would decrease it? Be sure to have your experiment and materials approved before you begin. Report your results.

INVESTIGATING SOME HUMAN REFLEXES

INTRODUCTION: Have you ever gotten something in your eye, and before you knew it your eye was tearing? Have you ever stepped on a sharp object, and by the time you realized it your foot was already off the object? These are common examples of a type of behavior known as a **reflex**. Reflex actions are automatic, involuntary, and occur very rapidly. A **stimulus** causes certain **neurons**, or nerve cells, to direct a specific **response** in a gland or muscle. The difference between this type of behavior and one in which you consciously, or voluntarily respond to a stimulus, lies in the fact that the **reflex arc**, or pathway taken by the nerve signals involved, by–passes the higher centers of the brain. No "thinking" or "awareness" is involved in a reflex action.

A reflex arc begins when a **sensory neuron** transmits an impulse from the **receptor** that detected the stimulus. The sensory neuron sends the "message" to the **central nervous systems** (CNS), usually the **spinal cord**. Within the spinal cord, an **interneuron** relays the impulse back out through a **motor neuron**. The motor neuron, in turn, directs the **effector**, either a muscle or gland, to respond.

Reflexes have both protective and adaptive functions. It is obviously beneficial to your skin to remove it from a hot stove before it has a chance to get seriously burned. The withdrawal reflex, therefore, is protective in nature. Adaptive reflexes help maintain the normal functioning of your body. For instance, when your eye focuses on a far away object, the lens has a particular shape. In order to change your focus and look at something close–up, the shape of your lens (and pupil) must automatically change. This action is an example of an adaptive reflex. In this investigation you will observe several human reflexes.

PURPOSE:

- What are reflex actions?

MATERIALS:

Flashlight or lamp
Clear plastic sheet
Crumpled piece of paper

PROCEDURE:

A. PUPILLARY REFLEX

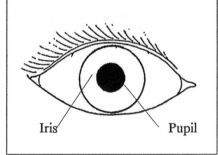

illus. 1

Work with another student (or stand in front of a mirror) to make observations.

1. Observe the size of the normal pupil in regular classroom light. (See illustration 1). On your data sheet, draw the appearance of the normal pupil.

2. Close your eyes and cover them with your hands for three minutes. Immediately upon opening your eyes, have your partner observe the size of your pupil. The pupil size will change quite rapidly, so look fast! Continue watching the size of the pupil until it returns to normal. Record your observations on the

data sheet.

3. Again close your eyes. Have your partner shine a flashlight at your left eye. Open your left eye and have your partner observe the changes that occur in the pupil as the light is shined on it. Record your observations.

B. ACCOMMODATION REFLEX

Accommodation is actually the process by which the curvature of the lens of the eye is changed in order to change focus. There is, however, a visible change in the pupil that accompanies the change in the lens shape. You will monitor the accommodation reflex by observing the pupil changes.

1. Look at and focus on some object across the room, or outside the window.

2. Have your partner observe the size of your pupil.

3. Now switch your focus to some printed material that you hold about 15 cm in front of your eyes. Your partner should carefully observe the changes in the pupil size that accompany your shift in focus. This activity may have to be repeated a few times to get valid observations, since the actual changes are rather small. Record your observations in Data Table 1.

C. BLINKING REFLEX

1. Hold a clear plastic sheet about 10 cm in front of your face.

2. Have your partner throw a crumpled piece of paper at the sheet while you are watching.

3. Your partner should observe a blinking reflex. Record the observations on the data sheet.

D. SKIN PINCH REFLEX

1. Stand facing your partner. Your partner should observe the size of your pupils.

2. Have your partner pinch the skin on the back of your neck, still watching the size of your pupils.

3. Record any changes observed in the pupil size.

E. STATIC EQUILIBRIUM (VESTIBULAR) REFLEXES

Static equilibrium refers to the orientation of the head or body in relation to the ground. It is maintained by the combined action of the vestibule of the (inner) ear, stretch receptors in the muscles of the neck and trunk, and sensory input from the eyes. Balance, or equilibrium, is controlled by a variety of reflex arcs, depending upon which sense organs are participating.

1. Stand upright with both feet together and your arms outstretched.

2. For the next two minutes, have your partner observe any body sway and corrective movements you make in order to maintain static equilibrium. Record your observations in Data Table 2.

3. Repeat the procedure standing on one foot. Record the observations.

4. Repeat steps 1 — 3 with your eyes closed. Record your observations.

DATA:

A.

PUPIL OBSERVATIONS

| NORMAL | AFTER DARK | AFTER BRIGHT LIGHT |

B.

DATA TABLE 1: ACCOMMODATION REFLEX OBSERVATIONS

Focus	Pupillary Changes
Far	
Near	

C. Blinking Reflex:

D. Skin Pinch Reflex:

E.

DATA TABLE 2: STATIC EQUILIBRIUM OBSERVATIONS

Standing Position	ABILITY TO STAND ERECT WITH EYES	
	Open	Closed
Both Feet		
One Foot		

CONCLUSIONS:

1. Why are all the actions observed in this investigation considered to be reflexes rather than conscious responses?

2. What is the biological significance of each of the reflexes observed? Are they protective, adaptive, or both?

3. Choose one of the reflexes studied, and trace its pathway from the stimulus to the response.

4. How does standing on both feet compare with standing on one foot in maintaining static equilibrium?

5. What is the importance of vision in maintaining static equilibrium?

SUGGESTIONS FOR FURTHER STUDY:

- Not only humans, but all animals with a complex nervous system have reflexes. Test for the scratch reflex in a dog in the following manner:

  Have a dog lie on its side or stand on a table. Gently scratch the dog's upper side just anterior to the tail, and lateral to the spine. Describe what happens. On which side was the response? Trace the reflex arc that produces this reflex.

- Some reflexes can be "overridden" by practice or conditioning. For example, a contact lens wearer soon becomes able to avoid the blink reflex that should occur when the cornea of the eye is touched. A circus performer on a high wire has trained himself to "ignore" a withdrawal reflex if he should step on a sharp object on the tightrope. Research how this "override" behavior can occur. Describe the pathways involved.

- Most human behavior is learned rather than reflexive. One type of learned behavior is a habit, which becomes ingrained to such an extent that it seems to be automatic and unconscious. Distinguish between a habit and a reflex. Describe the processes involved in habit formation. Do both benefit the person equally? How can a habit be more advantageous than a reflex? Less?

ANATOMY OF A REPRESENTATIVE VERTEBRATE

INTRODUCTION: Dissection of the frog has been a common activity in biology laboratories for decades. The primary reason for the frog's longevity in this area is that it is a **vertebrate** just like you. As such, most of its internal organs, their functions and relationships, are the same. The frog is large enough to be able to see its organs clearly, but small enough to be easily cared for in the laboratory. Frogs also are readily available and relatively inexpensive.

Before proceeding with the dissection, however, there are a few anatomical terms with which you must become familiar. These terms are among many that are used to indicate a relative position in anatomical studies. **Dorsal** and **ventral** are opposing surfaces that refer to the back and belly (front) areas, respectively. For instance, a shark's dorsal fin is attached to its back. Your belly–button is located on your ventral surface. **Anterior** and **posterior** are another pair of opposite terms. The former refers to the front, or head region, while the latter indicates the back–end, or tail region. As an example, your head is anterior to your chest, while your toes are posterior (or distal) to your ankles. Lastly for our purposes here, **lateral** means "on the side," whereas **medial** means toward the midline of the body.

As you investigate the internal anatomy of the frog, you will observe many organs. Remember that these structures are not just thrown into the body cavity in any which way. There is an organization to the body plan, and the relationships among the various organs and systems contribute to their effective functioning. As you identify various structures, keep in mind the following questions:

To which system does it belong?

What is its main function?

How is it connected to other organs in the system?

How does its form and structure relate to its function?

PURPOSE:

- To investigate the internal structure of a representative vertebrate, the frog.

MATERIALS:

Preserved frog	Dissecting pan
Scalpel	Dissecting pins
Scissors	Hand lens
Dissecting (teasing) needles	Forceps

PROCEDURE:

Obtain a preserved frog that has been rinsed thoroughly in tap water. Place the frog, ventral side up, on a dissecting pan. Insert pins, on an angle, into the upper and lower **appendages** (arms and legs) in order to anchor the frog in place. The frog's skin is attached loosely to its body. Pinch the skin along the midline of the abdomen, and lift it away from the body. With your scissors, cut the skin away so that all the muscles of the abdomen, upper part of the body, and upper part of the legs are exposed. (See illustration 1). Examine the underside of the frog's skin.

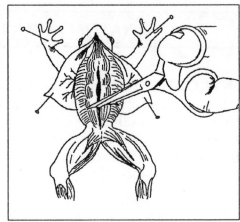

illus. 1

1. What do you notice on the inner surface of the skin? What might be the function of these structures?

    Look at the muscles of the frog's torso. Imagine how each set of muscles would contract as the frog moves.

2. Briefly describe the texture and direction of the various muscles you see.

You are now going to cut through the muscles and into the body cavity. Begin your incision by pinching the abdominal muscle, just anterior to the legs, and making a small "notch" cut. Insert the blunt end of your scissor into the cut, pointing in an anterior direction. **Do not cut too deeply, or you will damage the internal organs. Keep your incision shallow so that only the muscle is cut.** Make a medial cut (just off to one side of the dark blood vessel running down the mid–line) from the crotch up toward the sternum (breastbone). When you reach the bone carefully slide the scissor under it. The heart is directly under this bone, so be careful as you cut. Open the scissors quite wide, and close the blades with one swift cut to sever the bone. Continue cutting up toward the jaw, until you reach the jaw bone. Do not cut any further. Now make horizontal cuts through the muscle, so that it may be peeled back to expose the internal organs. (See illustration 2). Pin the muscles to the tray in order to keep them out of your way.

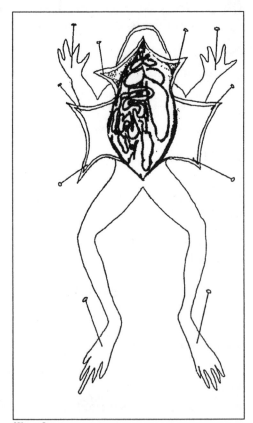

illus. 2

If your frog is a female, A large mass of black eggs may be hiding many of the internal organs. Observe the egg mass, then carefully remove it from the body cavity using your forceps and scissors.

3. What do you notice about the number of eggs produced by a female frog? Why do you think that so many eggs are required?

Now you will locate and observe many of the major organs of the frog. On a separate sheet of paper, sketch each of the structures observed, maintaining their relative size.

Observe the frog's **heart,** located anteriorly in the thorax (neck region). It is found within a triangular pericardial sac, or membrane. Cut open the pericardium so that you can see the three chambers of the heart. The frog has two upper atria to receive blood, and one lower, pointed ventricle which pumps blood.

4. In what way is the frog's heart different from yours? Is this more or less efficient?

Locate the large artery that leaves the heart. This is the aorta.

5. What type of blood is carried in the frog's aorta? Where does the blood go that travels in this artery?

The largest internal organ of the frog (and human) is the reddish **liver**. There are three lobes to the liver, connected to each other beneath the heart. Examine the liver. Carefully lift up the right lobe to see the greenish **gall bladder**, which is attached to the underside of the liver. This sac–like organ stores bile, a digestive chemical that emulsifies fat. With a hand lens, see if you can locate the thin, white bile duct that leads from the gall bladder.

6. Where does the bile duct go? Is there any difference in the structure or function of the liver and gall bladder of the frog and human? Explain.

Once you have observed the liver and gall bladder, remove them from the body cavity with your forceps and scissors.

The **alimentary canal** is composed of the esophagus, stomach, small intestine and large intestine. This organ system is responsible for the intake and digestion of food and the elimination of wastes. First locate the whitish, curved, sac–like **stomach**, lying near the middle of the abdomen towards the frog's left side. Anterior to the stomach is the **esophagus**, or food tube. Trace the esophagus as far forward as you can. Posterior from the stomach is the **small intestine,** a long tube that is tightly wrapped and folded within a protective membrane. Further posterior along the alimentary canal is the **large intestine**, much shorter and thicker than the small intestine. Trace the small intestine to where it leads into the large. The large intestine leads deep into the posterior abdominal cavity, ending at the **anus**, between the frog's hindlimbs. Near the anus you should also see the two–lobed **urinary bladder**, which joins with the large intestine. This sac stores urine until it is removed from the frog's body. Remove the alimentary canal, in one piece if possible, from the frog's body. Holding the stomach with forceps, snip the esophagus at its anterior end. As you remove the canal, snip away the membrane that holds the small intestine, so that it can uncoil and you can see its true length. Notice the size of all the organs. Observe the numerous blood vessels in the membrane of the small intestine.

7. What do you think is the function of the numerous blood vessels in the membrane of the small intestine?

Also in the membrane, about mid–way between the stomach and the large intestine, is a small, spherical, dark–red **spleen**.

8. Is the spleen a part of the digestive system? What is its function?

Cut open a section of the stomach and small intestine. Observe the inner surface of these organs with a hand lens.

9. Describe the inner surface of the stomach and small intestine. Can you see any villi? What is the purpose of this type of surface?

Your frog may have large, yellow, finger–shaped organs in its body cavity. These are **fat bodies**, in which extra food is stored to be used during hibernation or gamete production. Remove these if they are present.

10. At what time of year would you expect to find large fat bodies in a frog? What tissue do humans have that is comparable to these structures?

On the dorsal surface at the anterior end of the body cavity, locate the two dark–reddish **lungs**. If the lungs are deflated (most likely), they will appear long and thin. If they are inflated, they will look like oval balloons. Look for the anterior connection of the lungs to the **larynx** and **glottis**.

11. Describe the structure and texture of the lungs. How is this an adaptation for their function?

Deep in the posterior abdominal cavity, on either side of the backbone, are two reddish–brown **kidneys**. These organs excrete urine from the frog's body, first into the urinary bladder, then out through the cloaca, or lower part of the large intestine. Trace the path of urine from the kidneys, through tubes, to the cloaca.

On the anterior ventral surface of each kidney you should see a yellowish mass of tissue. These are the **adrenal glands**, and serve the same function in frogs as they do in humans.

Locate the **backbone** on the dorsal surface of the frog. Each of the bones is called a **vertebra**. Within the vertebral column is the spinal cord. Spinal nerves come from the spinal cord out to various parts of the frog's (and human's) body. See if you can locate one or more of the ten pairs of spinal nerves. (The sciatic nerve is the largest of these, and leads to the muscles in the thighs).

Unless your teacher directs you otherwise, clean up your dissection now by wrapping the frog, and any removed parts, in paper towels. Throw the paper towels away according to the directions of your teacher. Wash off your dissecting instruments and tray.

DATA:

1._____

_____

2._____

_____

3._____

_____

4._____

5._____

_____

6._____

_____

7._____

_____

8._____

_____

9._____

_____

10._____

_____

11._____

_____

CONCLUSIONS:

Base your answers to the following questions on the organs of the frog listed below:

heart	liver	gall bladder
esophagus	stomach	small intestine
large intestine	lungs	glottis
spleen	kidneys	adrenal glands
fat bodies	vertebral column	spinal cord

1. For each organ, identify the system to which it belongs, and its major function.

2. For each organ, determine if there is an equivalent structure in your body. Is it relatively in the same location?

SUGGESTIONS FOR FURTHER STUDY:

- This activity did not include any details of the muscular, skeletal or nervous systems of the frog. These systems are worthy of further study. Obtain and wash a fresh, preserved frog.

    1. Remove all the skin of the frog and expose the muscles. Consulting charts and reference texts, compare the muscular system of the frog with that of the human.

    2. Remove all the tissue from a frog in order to obtain the skeleton. This can be accomplished by soaking the skeleton, after you have cut away as much tissue as possible, in clorox bleach. Be sure to do this in a well ventilated area. Do not leave the skeleton in the bleach any longer than necessary to remove the remaining tissue. Mount the skeleton and identify the bones. Compare the skeletal system of the frog to that of the human.

    3. Observe the frog's brain. Place a frog on a dissecting tray with the dorsal surface up. Remove the skin from the skull (between the eyes). With a scalpel, shave or whittle the bone of the skull between the eyes. Use very shallow strokes, peeling off the bone in layers or shavings. When the bone becomes thin, peel it off very carefully using forceps. Be sure not to dig the points of the forceps into the brain tissue. Using charts and reference texts, identify the parts of the frog's brain, and compare them with those of the human.

WHAT ARE THE RAW MATERIALS OF PHOTOSYNTHESIS?

INTRODUCTION: Autotrophs are organisms that produce their own food from inorganic raw materials. Green plants are by far the most common autotrophs, producing their food through the process of **photosynthesis**. Look at the equation below, which summarizes the photosynthetic process.

$$6CO_2 + 12H_2O \xrightarrow[\text{LIGHT}]{\text{CHLOROPHYLL}} C_6H_{12}O_6 + 6O_2 + 6H_2O$$

This equation was developed through the work of many scientists who carefully studied the chemical processes involved in photosynthesis. In this activity, you will verify the necessity of three of the raw materials identified in this equation, **carbon dioxide, chlorophyll**, and light.

PURPOSE:

- Are carbon dioxide, light and chlorophyll necessary for photosynthesis?

MATERIALS:

4 sprigs of Elodea	Bunsen burner
4 test tubes with stoppers	2 geranium plants
Test tube rack	1 Coleus plant
Straw	Boiling water bath
Bromthymol blue solution	Alcohol bath
Lugol's iodine solution	Scissors
Test Tube holder	Forceps
Goggles	

PROCEDURE:

Part I: IS CARBON DIOXIDE NEEDED FOR PHOTOSYNTHESIS?

Aquatic plants, such as Elodea, carry out photosynthesis under water, and will be used in this investigation. Carbon dioxide may be added to a sample by exhaling your breath, through a straw, into a test tube of water. The presence of carbon dioxide is confirmed by the indicator bromthymol blue.

1. As a preliminary test, half–fill two test tubes with water. Add several drops of bromthymol blue to each tube. Into one of the tubes, add carbon dioxide by gently blowing through a straw until you see a color change. Record the results in Data Table 1. From the other tube, remove any carbon dioxide by heating it gently over the flame of a Bunsen burner for several minutes. **Hold the test tube with a test tube holder, and point the open end of the tube away from you. Wear goggles when working with an open flame.** Record the color of this tube on the data table.

2. Now set up your experiment to determine if an Elodea plant uses carbon dioxide during photosynthesis. Place four test tubes in a rack, and label them 1–4. Into Tube 1, place water, carbon dioxide, a few drops of bromthymol blue, and a healthy sprig of Elodea. Tube 2 will be a control *without* the Elodea. Tube 3 will contain the same materials as Tube 1, but the carbon

dioxide should be removed before adding the Elodea. (Allow the water to return to room temperature before adding the plant). Prepare tube 4 as a control for 3 by omitting the plant from the set–up. Stopper all tubes.

3. Record the color of the bromthymol blue in each of the four tubes on Data Table 2.

4. Place the test tube rack near a window in bright sunlight, or under a strong lamp for 24–48 hours (or until you see a color change in at least one of the tubes). Record the final colors of all tubes in the data table.

Part II: IS LIGHT NEEDED FOR PHOTOSYNTHESIS?

In order to determine if photosynthesis has occurred, it is a common practice to test for the presence of starch in the leaves of the plant. The reason for this is that the sugars produced by the leaf are quickly synthesized into starch for more convenient storage.

1. Place a healthy geranium plant in a dark area, such as a cabinet, for 48 hours prior to testing. Leave another plant of the same type in sunlight for the same amount of time.

2. After 48 hours, remove one healthy leaf from each of the plants. In order to identify which leaf was in the light and which was in the dark, cut a small "V" shaped notch in the leaf from the plant in the dark.

3. To test for the presence of starch, the green pigment, chlorophyll, must first be removed from the leaf. Why is this necessary? In order to remove the chlorophyll, the leaf must first be softened, and the cells killed by placing them in boiling water for five minutes. Set up a boiling water bath by adding water to a large beaker and heating it on a hot plate. Drop the leaves into the water bath, and remove them, after five minutes, with forceps. **Wear goggles.**

4. Describe the condition of the leaves after removal from the boiling water. Has the chlorophyll been removed? What does this tell you about the solubility of chlorophyll?

5. In order to dissolve the chlorophyll, the leaves must be heated in alcohol, an organic solvent. Since alcohol is flammable, you should never heat it directly over a flame. A safe way in which to heat the alcohol is to put it into a small beaker (approximately 200 ml), and place this beaker inside a larger beaker (1000 ml) which is partially filled with water. Place the larger beaker on a hot plate and heat it until the water boils. Place the leaves into the hot alcohol for 5–10 minutes, or until all the green color has been removed. **Wear goggles when heating the alcohol. Do not let the alcohol boil over the rim of the beaker.**

6. Remove the bleached leaves and dry them on paper towels. Test each leaf for starch by flooding it with Lugol's iodine solution. (A blue–black color indicates the presence of starch). Record your results in Data Table 3.

Part III: IS CHLOROPHYLL NECESSARY FOR PHOTOSYNTHESIS?

Variagated leaves are those that do not have a uniform distribution of chlorophyll. By comparing the pattern of chlorophyll distribution in a variagated leaf to the pattern of starch deposits, it is possible to determine if chlorophyll is needed for photosynthesis.

1. Remove a healthy leaf from a Coleus plant that has been in the sun for several days. In the data section, make an accurate sketch of your leaf, indicating where chlorophyll is present by shading your sketch with pencil.

2. Using the same procedures as you did in part II, remove the chlorophyll from your leaf.

3. Test the leaf for the presence of starch as before. Record your results by making another sketch of the leaf, shading in the areas in which you found starch to be present.

DATA:

Part I:

DATA TABLE 1: PRELIMINARY TEST OF BROMTHYMOL BLUE INDICATOR

Contents of Test Tube	Carbon Dioxide Water	Water Without Carbon Dioxide
Color of Bromthymol Blue		

DATA TABLE 2: RESULTS OF CARBON DIOXIDE EXPERIMENT

Test Tube Contents	1 Elodea + CO$_2$	2 CO$_2$ Only	3 Elodea without CO$_2$	4 Water Only
Color at Start				
Color after 48 Hours				

Part II:

3._____

4._____

_____

DATA TABLE 3: RESULTS OF LIGHT EXPERIMENT

Light Condition	Light	Dark
Presence of Starch		

Part III:

1. Leaf sketch showing areas of chlorophyll

3. Leaf sketch showing areas of starch

CONCLUSIONS:

1. Refer to your results as shown in Data Table 2. In which tube(s) was carbon dioxide used? How can you tell?

2. What conclusion can you draw from your answer to question 1?

3. Did any changes occur in tube 3 in Part I? If so, how can you account for these?

4. What was the purpose of tubes 2 and 4 in Part I?

5. Based on your results from Part II, is light needed for photosynthesis? Why or why not?

6. Based on your results from Part III, what is the relationship between starch production and the presence of chlorophyll in a leaf?

7. Based on your answer to question 6, is chlorophyll required for photosynthesis?

SUGGESTIONS FOR FURTHER STUDY:

- Design and conduct an experiment to determine the effect of different wavelengths (colors) of light on the process of photosynthesis. Obtain teacher approval, and then conduct your experiment. Report your results.

- Design and conduct an experiment to determine the effect of different intensities of light on the process of photosynthesis. Obtain teacher approval, and then conduct your experiment. Report your results.

WHAT ARE THE PRODUCTS OF PHOTOSYNTHESIS?

INTRODUCTION: Photosynthesis, the process by which green plants synthesize food from inorganic raw materials, is the basis for almost all food production on our planet. As a result of this process, **oxygen** is given off as a waste product. It is apparent, therefore, that without green plants and their ability to carry out photosynthesis, life as we know it would not be possible on Earth. In this exploration you will verify the products of photosynthesis, as illustrated by the equation below.

$$6CO_2 \; + \; 12H_2O \; \xrightarrow[\text{Light}]{\text{Chlorophyll}} \; C_6H_{12}O_6 \; + \; 6O_2 \; + \; 6H_2O$$

PURPOSE:

- What are the products of photosynthesis?

MATERIALS:

Several leaves from two different plants	Pinch of sodium bicarbonate
Mortar and pestle	Wooden splint
Pinch of fine sand	Scalpel
Stirring rod	4 rubber stoppers
Benedict's solution in a dropping bottle	Hot plate
2 test tubes and a test tube holder	Ring stand
Boiling water bath	Clamp
Alcohol bath	Elodea sprigs
Short-necked funnel	1000 ml beaker
Lugol's iodine solution in a dropping bottle	Goggles
	Forceps
	Lamp

PROCEDURE:

Part I: WHAT FOOD IS PRODUCED THROUGH PHOTOSYNTHESIS?

According to the equation above, plants should produce **glucose** as a result of photosynthesis. Different types of plants produce glucose in varying amounts. Glucose, however, is quickly converted into **starch** for easier storage in the leaves, and other organs of the plant. In this section, you will test for both sugar and starch production.

1. Remove several leaves from a tree, bush, or houseplant. On Data Table 1, identify the type of plant you selected. Put one leaf aside to use later in this investigation.

2. Shred the leaves into a mortar, add a pinch of fine sand, and grind them with a pestle. Add a few milliliters of warm (approximately $35^{\circ}C$) water, and stir the leaf mixture for several minutes.

3. Pour the leaf–water mixture into a test tube until it is approximately one-quarter filled. Add an equal amount of Benedict's solution to the tube and mix.

4. Using a test tube holder, carefully place the tube into a boiling water bath. **Wear goggles when working with boiling liquids.**

5. After 5 minutes, carefully remove the test tube and observe the color of the mixture. Benedict's solution is an indicator for sugar. A green color indicates less than 1% sugar. A yellow–orange color indicates a sugar concentration of 1–2%, while a brick red color denotes a concentration of over 2%. No change in the color of the Benedict's solution is a negative result. Record the approximate percent of sugar in your leaf sample in Data Table 1.

6. Repeat the above procedures using another type of leaf. As before, retain one of these leaves to use later. Record your data in the Data Table.

In order to test for the presence of starch, the chlorophyll must first be removed from the leaf. Since chlorophyll is not soluble in water, it must be removed by an organic solvent, such as alcohol. For this test, you will use the two leaves that were saved from steps 1 and 6.

7. The leaves must first be killed and softened before the chlorophyll can be removed. To accomplish this, place the leaves into a boiling water bath. After 5 minutes, carefully remove them with forceps.

8. Set up an alcohol bath by adding alcohol to a small beaker (200 ml) and placing this beaker *inside* the beaker containing the boiling water. **Hold the beaker with tongs, and lower it gently into the boiling water. Wear goggles. Alcohol is flammable, be extremely careful during this activity.** Place your wilted leaves in the alcohol bath for 10 minutes, or until the chlorophyll is completely bleached out.

9. Using forceps, remove the leaves from the alcohol bath, and dry them on paper towels. Test each leaf for the presence of starch by flooding it with Lugol's iodine solution. A blue–black color is a positive test for starch. Record your results in the data table.

Part II: IS OXYGEN PRODUCED DURING PHOTOSYNTHESIS?

In this part of the investigation you will use an aquatic plant, since it is easier to detect the presence of a gas when it is collected in water.

1. Select 4–5 healthy sprigs of Elodea, and carefully cut off 1–2 mm of each stem with a sharp scalpel or single–edged razor blade. Place the sprigs in the bowl of a short–stem funnel, with the cut ends pointing up into the neck.

2. Fill a 1000 ml beaker with water, and add a pinch of sodium bicarbonate. (This will supply carbon dioxide to the photosynthesizing plant). Place four rubber or cork stoppers in the bottom of the beaker, and lower the funnel into the beaker so that the bowl rests on the stoppers. (See the illustration on the next page.)

3. Fill a test tube with water. Holding your thumb over the open end of the tube, invert it over the neck of the funnel. Take care not to lose any of the water

from the tube when you lower it over the funnel. Be sure the open end of the tube is *under* the water in the beaker. Hold the tube in place by clamping it to a ring stand.

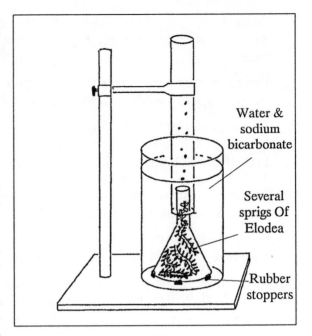

Water & sodium bicarbonate

Several sprigs Of Elodea

Rubber stoppers

4. Position a bright lamp so that its light shines on the plants in the beaker. Observe your experimental set–up over the next 24 hours, or until all the water in the test tube has been displaced by gas. Record any observations you make on Data Table 2.

5. Once you have collected the gas given off by the Elodea, you can test it to determine if, in fact, it is **oxygen**. Since oxygen supports combustion, a glowing splint should burst into flame when placed in the test tube. Carefully remove the test tube from your apparatus and quickly turn it right–side up. Why doesn't the oxygen gas escape from the tube?

6. Light a wooden splint, and blow out the flame once it begins to burn. Immediately thrust the glowing end into the test tube. Record your observations.

DATA:

Part I

DATA TABLE 1: WHAT FOOD IS PRODUCED DURING PHOTOSYNTHESIS?

Leaf Type	Benedict's Results	Starch Results (+ or -)

Part II

DATA TABLE 2: RESULTS OF GAS PRODUCTION DURING PHOTOSYNTHESIS

Observations	Splint Test

5._____

CONCLUSIONS:

1. According to your results, what are the products of plant photosynthesis?

2. What factors, other than species of plant, might account for the difference in sugar concentrations between your two leaf samples?

3. Why must the chlorophyll be removed from a leaf before it can be tested for starch?

4. In Part II, why was the funnel supported by stoppers, instead of resting directly on the bottom of the beaker?

SUGGESTIONS FOR FURTHER STUDY:

-   Design and conduct an experiment to test the effect of varying temperatures on a plant's production of sugar or oxygen. Have your experiment approved by your teacher before proceeding. Report your results.

-   Design and conduct an experiment to measure the rate of oxygen production during photosynthesis. Have your experiment approved, and report your results.

-   It has been reported that a large portion of the world's oxygen supply is produced by plants of tropical rain forests throughout the world. Recently, these rain forests have been destroyed by encroaching civilization. Research the environmental effect of rain forest destruction as it relates to oxygen supply. Excellent resource material may be found by contacting state, national, and international environmental agencies. Prepare a paper to report your findings.

CLASSIFICATION AND IDENTIFICATION OF LEAVES

INTRODUCTION: Collecting and identifying leaves is an enjoyable and educational activity. Plant collections, like stamps or coins, can grow into an exciting pastime as your knowledge and expertise in this area expands. With a fundamental knowledge of a few characteristics of leaves, and a good field guide for reference, you will soon be able to identify the specimens you collect. Your collection will also make an attractive and informative display when mounted.

PURPOSE:

- How can trees be identified from their leaves?

MATERIALS:

Field guide to trees
White mounting paper or oak tag
Rubber (Duco) cement or clear contact paper
Newspaper
Plant press or heavy books

PROCEDURE:

You are going to prepare a collection of at least 10 different leaves. The leaves should be collected from local trees (or shrubs if your teacher approves), and you will need 2 typical specimens from each tree. As you collect your specimens, place them between the pages of a book or magazine, where they can be spread out and flattened. Indicate the location from which your specimen was taken, and the type of tree (if known). Of your 10 samples, 3 should be from evergreen and 7 from deciduous trees. Your teacher will inform you of any modification in this plan. Once you have collected all your specimens, you must mount them and identify them. Follow the directions below for each of these procedures.

MOUNTING LEAVES:

Several methods for mounting leaf specimens are available. The one here may be modified or substituted with approval from your teacher. Place your two leaves from each tree between several pieces of newspaper. Place the paper under a phone book, or several heavy books, or in the middle of a plant press, for about 5 days. This will dry out the leaves before mounting. When they are removed, place them side by side on a sheet of heavy white paper. You may glue your leaves onto the paper using rubber (Duco) cement, or cover them with clear contact paper. Be sure each leaf is flat. Place one of the leaves so that the upper surface is showing, and the other so that the lower surface shows. When you identify your specimens, write both the scientific and common names below each sample. Be sure to include your name, and the date and location of the sample on each page.

IDENTIFICATION OF LEAVES

In order to correctly identify the tree from which your leaf was taken, the following terminology and characteristics should be reviewed:

Leaves may be divided into two general categories – (see illustration 1)

    A. Needlelike or Scalelike
    B. Broad–leaved
       1. Compound
       2. Simple

Needle or scalelike leaves are commonly found in **coniferous** (cone–bearing) trees, such as the pine. These thin, pointed leaves are found in clusters of 2–5 per bunch, and generally remain green all year long (evergreens).

Broad–leaved trees are usually **deciduous**, meaning their leaves are shed in the winter. Broad–leaved trees are the ones most of us think of when we visualize a "typical tree."

Compound leaves are usually divided into 3 or more **leaflets**, which are attached to a non–woody stalk. This stalk, in turn, is attached to the woody twig of the tree branch. **Simple leaves** consist of a single leaf **blade** attached directly to the twig. Both compound and simple leaves may vary in certain characteristics which can be used to help identify them. These characteristics include: leaf shape, types of edges (margins), number of lobes, pattern of veins, surface quality, shape of tip. Some of these may be seen in illustration 2. Refer to your field guide for other identifying characteristics.

Using one or more field guides, or other reference material, identify the tree from which each of your leaf samples was taken. Write its common and scientific name below each mounted specimen.

Illustration 1: Leaf Categories

BROAD–LEAVED

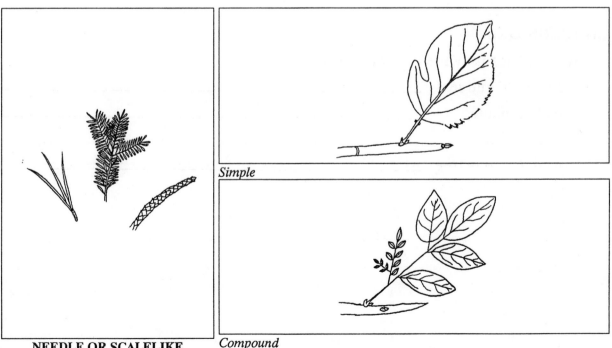

Simple

NEEDLE OR SCALELIKE   *Compound*

CHARACTERISTICS OF LEAVES

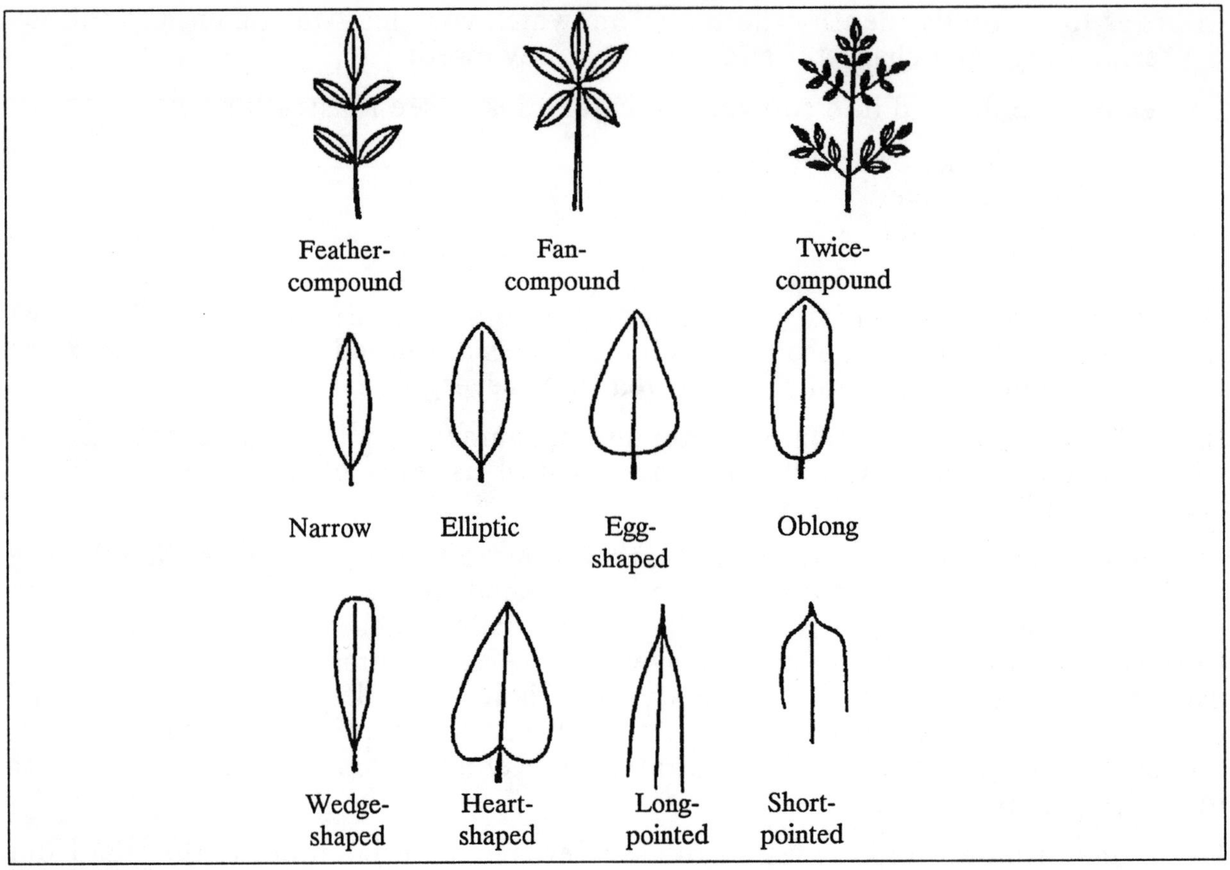

illus. 2

CONCLUSIONS:

1. What is meant by the following terms:

    a) compound leaf _____

    b) leaf blade _____

    c) leaf vein _____

    d) deciduous tree _____

SUGGESTIONS FOR FURTHER STUDY:

- When identifying unknown trees, more characteristics than those of leaves are usually examined. Using an appropriate field guide or other reference, investigate the other information usually collected when trying to make accurate identifications. Using these added criteria, identify at least 10 other local trees or shrubs. Mount and label your specimens appropriately. Display your collection in the classroom.

THE EFFECTS OF AUXIN ON PLANT GROWTH

INTRODUCTION: Auxins are a type of hormone which regulate growth and behavior in plants. These compounds, one example of which is **indoleacetic acid (IAA)**, are produced in many growing regions of a plant, such as root and stem tips, developing leaves, flowers and fruits. Some of the hormones cause cell elongation while others result in cell division. If auxins are distributed around the plant unevenly, differential growth – such as bending or "turning" – will result. **Tropisms**, or plant growth responses to stimuli, are controlled by this unequal distribution of certain auxins. Auxins are also known to cause flowers to bloom, fruits to ripen and leaves and roots to grow. In this activity, you will investigate the effects of the auxin IAA on the directional growth of bean seedlings.

PURPOSE:

- What effect does auxin have on the directional growth of bean seedlings?

MATERIALS:

9, 2–4 week old bean seedlings
100 g of lanolin paste
100 mg of IAA (indoleacetic acid)
2 ml of absolute ethanol

PROCEDURE:

Prepare a lanolin paste of IAA according to the following directions:

1. Dissolve 100 mg of IAA in 2 ml of absolute ethanol.

2. Mix the solution thoroughly with 100 g of lanolin.

To prepare your experimental plants, apply equal quantities of your lanolin–IAA paste in each of the locations shown on the illustration.

Prepare two sets of control plants. To one set of three plants, do nothing. To another set of three plants apply lanolin, only, in the same locations as you did for the experimental plants. Label each of your plants appropriately.

Place the plants in the same maximum light conditions in your classroom. Observe them each day for the next week. Record your observations on the Data Table.

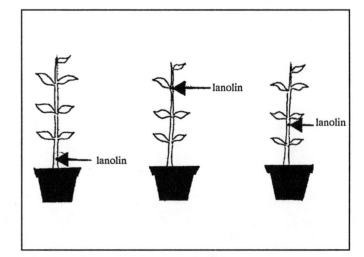

DATA:

DATA TABLE: SEEDLING GROWTH WITH AND WITHOUT IAA

Day	Experimental Plants	Plants Treated With Only Lanolin	Plants Receiving No Treatment
1			
2			
3			
4			
5			
6			
7			

CONCLUSIONS:

1. Why did you prepare three seedlings with lanolin only?

2. Summarize the effects of IAA on the directional growth of seedlings.

3. Offer an explanation as to why the results, as described in your answer to question 2, occurred.

SUGGESTIONS FOR FURTHER STUDY:

- Design and conduct an experiment to determine the effect of indoleacetic acid on the growth of roots. Have your experimental design approved by your teacher before you begin. Report on your results.

- Synthetic auxins may be used as weed killers, the most common being 2,4–dichlorophenoxyacetic acid (2,4–D). In general this and other related compounds kill broad–leaved plants without harming monocots (grasses and grains). This selective property of 2,4–D has led to its extensive use in fields, lawns, and golf courses. Common weeds, such as wild mustard, dandelion and plantain, can be killed without any harm to the grasses or crops. If 2,4–D is still available in your area, design and conduct an experiment to show the selective herbicidal action of this auxin. Using concentrations of .01%, .1% and 1.0% in lanolin paste, compare the effects on rye and bean seedlings. **Use care when handling this chemical. See your teacher for appropriate safety measures.** If 2,4–D has been removed from the market in your area, find out why. What were the problems, if any, with this compound? Report on your findings.

WHAT IS THE NORMAL TIME SEQUENCE OF MITOSIS?

INTRODUCTION: Mitosis is an almost universal process by which plants and animals increase their number of cells. Once begun, mitosis is usually a continuous process composed of two basic stages. Nuclear division, or **karyokinesis**, is generally followed by division of the cell's cytoplasm, or **cytokinesis**. During nuclear division, **chromosomes** are replicated and two identical nuclei are reformed. The cytoplasm generally divides between these two nuclei, but not always in equal amounts.

Even though the entire process is a continuous one, karyokinesis is arbitrarily divided into five major steps for ease of study. In each of these stages, characteristic changes may be seen in the nuclear material.

In this investigation, you will look at cells of a root tip, the rapidly growing area of a plant root. These cells will all be in various stages of mitosis. By counting the numbers of cells in each stage, it is possible to extrapolate the actual time needed to complete each phase. In general, the more cells you find in a specific stage of mitosis, the longer that stage takes to complete. Fewer cells in a given phase indicate a shorter period of time for completion.

Before you begin your investigation, familiarize yourself with the names and characteristic appearance of plant cells in each phase. A summary is given below. (See illustration 1.)

Interphase: Although this is usually not considered an active phase of mitosis, important changes are occurring in the cell's nucleus. The appearance of the cell is the same as a "typical," non–dividing cell. On a molecular level, however, chromosomes are replicating in preparation for nuclear division.

Prophase: Usually considered the first stage of active mitosis, prophase includes the disintegration of the **nuclear membrane** and the **nucleolus**. Also during this phase, the chromatin material shortens and thickens to become visible strands. **Spindle fibers** form, and by attaching to each replicated chromosome, help them migrate towards the center of the cell.

Metaphase: This phase is recognized as the time at which all the chromosomes have lined–up on the equitorial plane (mid–line) of the cell.

Anaphase: Each **chromatid** pair separates and is pulled to opposite ends of the cell by the spindle fibers. Each chromatid is now considered an entire chromosome. By the conclusion of this stage, two identical groups of chromosomes have migrated to either pole of the cell.

Telophase: During the last stage of mitosis, the chromosomes reform into interphase condition. Nuclear membranes and nucleoli reform around each group of chromosomes. The chromosomes themselves become long and indistinct. Spindle fibers disintegrate. Cell division usually occurs during telophase. In plant cells, a new cell wall, called a **cell plate**, forms along the equitorial plane. By the end of telophase, two identical **daughter cells** are formed.

ILLUSTRATION 1: STAGES OF MITOSIS

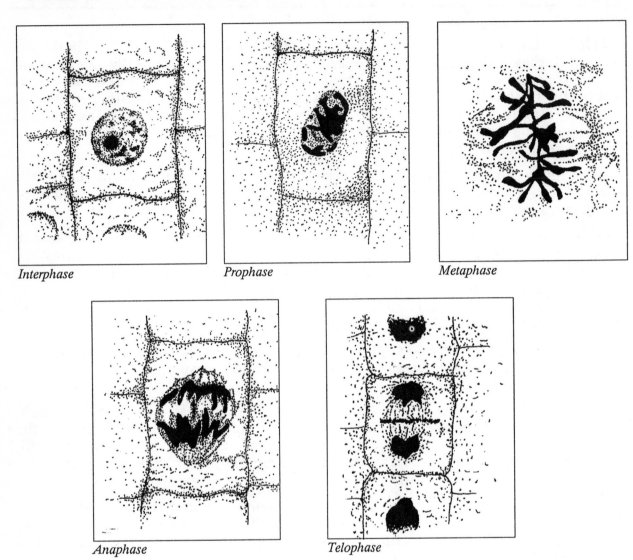

Interphase   *Prophase*   *Metaphase*

Anaphase   *Telophase*

PURPOSE:

- What is the normal time sequence for the various stages of mitosis and how is this timing altered in cancer cells?

MATERIALS:

Prepared slide of onion root tip
Compound microscope
Immersion oil (if applicable)
Lens paper
Graph paper

PROCEDURE:

Part I: NORMAL MITOTIC DIVISION

1. Obtain a microscope and onion root tip slide. Clean both before proceeding.

2. Scan the slide under **low power** first and locate the region of rapidly growing cells directly above the root cap. (See illustration 2.)

3. Switch to **high power** and center your slide so that you have a field of view in which all the cells are in various stages of mitosis (including interphase). Be sure to adjust your light for optimum viewing.

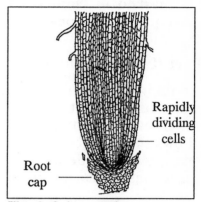

Illus. 2: Onion root tip

4. You are now going to identify the stage of **each** cell in your field. Starting at the top right corner of the field, record the stage of each cell in Data Table 1. Count your cells in a systematic manner. This will be considered area 1.

5. After completing the count in the first area, move your slide to a new area and perform the identification and count a second time. Record this data in the table as well.

6. Repeat the procedure a third time, with yet another field of view.

7. On a separate sheet of plain, unlined paper, make a drawing of one cell in each of the various stages. Be sure to draw only what you actually see, but include all details that are visible. Your drawings will not necessarily look exactly like the ones in illustration 1.

8. Return to Data Table 1 and record the total numbers of cells in each phase, for all three trial areas.

9. Add the total number of cells viewed in each stage. Write the total count of cells viewed in all three trials in the appropriate place on the data table.

There is a direct relationship between the number of cells counted in a given stage of mitosis and the time that that stage takes to complete. This may be calculated if the total time for mitosis in onion root tip cells is known. (The total time is measured from interphase to interphase). It is generally acknowledged that this time for onion cells is 720 minutes (12 hours). Set–up a ratio of the number of cells in each phase, compared to the total number of cells counted. Then multiply this fraction by the total time (720 minutes) needed to complete one mitotic division. Thus, the time for a specific phase is equal to:

$$\frac{\textbf{number of cells in a specific phase}}{\textbf{total number of cells counted}} \quad \textbf{X} \quad \textbf{720 min} \quad = \quad \textbf{time for specific phase}$$

10. Using your data, calculate the time required for the completion of each stage. Be sure to use the totals for all three trials. Enter these results in the appropriate column of Data Table 1.

11. Prepare a **bar graph** to illustrate your results. The vertical axis should be marked in "minutes to complete each stage." On the horizontal axis, allow equal space for each of the stages, beginning with interphase and ending with telophase.

Part II: MITOTIC DIVISION IN CANCER CELLS

An important characteristic of cancer cells, is that they no longer follow their normal timing of mitosis. You may have heard of cancer cells being "runaway" cells which have no controls on their rate of reproduction. It is this characteristic that allows some cancer cells to grow and spread quite rapidly. In this section, you will analyze data to determine the differences in timing of mitosis between normal and cancerous stomach cells of the chicken.

1. Study the data in Table 2. Assume that the total time needed for one normal mitotic division of these cells is **625** minutes. Calculate, in the same manner as before, the total time needed for each normal phase of mitosis. Enter this data in the appropriate column of Data Table 2.

2. Repeat this analysis for the data in Table 3. In the case of cancer cells, however, the total time needed for one mitotic division is only **448 minutes**. Enter the time required for each stage in Data Table 3.

DATA:

DATA TABLE 1: COUNT AND TIMING OF CELLS IN VARIOUS STAGES OF MITOSIS

Stage of Mitosis	# Cells in Area 1	# Cells in Area 2	# Cells in Area 3	Total # Cells	Time in Minutes
INTERPHASE					
PROPHASE					
METAPHASE					
ANAPHASE					
TELOPHASE					
			TOTAL CELLS COUNTED		720	TOTAL TIME IN MINUTES

DATA TABLE 2: NORMAL CHICKEN STOMACH

Stage of Mitosis	Total # Cells	Time In Minutes
INTERPHASE	440	
PROPHASE	40	
METAPHASE	8	
ANAPHASE	2	
TELOPHASE	10	
TOTALS	500	625

DATA TABLE 3: CANCER CHICKEN STOMACH

Stage of Mitosis	Total # Cells	Time in Minutes
INTERPHASE	424	
PROPHASE	50	
METAPHASE	12	
ANAPHASE	3	
TELOPHASE	11	
TOTALS	500	448

3. Prepare another bar graph, similar to your first, using the data from Tables 2 and 3. Put the data for both the normal and cancer cells in each phase directly next to each other.

CONCLUSIONS:

Referring to your data and graphs from Part I, answer the following:

1. Which stage in the mitotic cycle takes the most time? What *percentage* of the total time is this?

2. Why do you think that this stage (the one in question 1) takes so long? What activities, in relation to mitosis, are occurring during this phase?

3. Which stage is the second longest? What *percentage* of the total time does this stage take up?

4. Again, what events are occurring during the stage identified in question 3?

5. List the remaining stages, in order, from longest to shortest duration.

Referring to your data and graphs from Part II answer the following:

6. How does the data for each phase in the *normal* chicken cell compare with that of the onion root tip cell? Are the *percentages* of time for the two longest phases similar? Can you make a general conclusion based on this information?

7. In which stages are the most dramatic differences in timing between normal and cancerous chicken cells?

8. What nuclear and cytoplasmic changes would you expect to find in cancer cells, as compared to their normal counterparts? (HINT: what events would be most affected by the alteration in the timing sequence of mitosis?)

SUGGESTIONS FOR FURTHER STUDY:

- Perform an experiment to determine at what time of day cell division is most active. In order to do this, you will have to prepare your own slides. Ask your teacher for instructions on the preparation of onion root slides using the "squash technique." Meanwhile, begin sprouting roots on five onion bulbs (see illustration). After 3 or 4 days, you will prepare one slide from each bulb, cut at 5 hour intervals throughout the day. Note the time of day that each slide is prepared. Observe your slides as you did in Part I of this investigation, recording the number of cells undergoing mitosis on three separate areas of each root tip. What conclusions can you draw regarding mitotic activity at different times of the day?

- Perform an experiment similar to the one above, using other bulbs, such as daffodil or hyacinth.

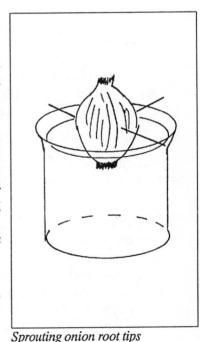

Sprouting onion root tips

CAN YOU CLONE A PLANT?

INTRODUCTION: Most multicellular plants can reproduce sexually, using flowers or cones to produce seeds. Many plants, however, also have the ability to reproduce asexually. Using parts that are not normally involved in reproduction, such as leaves, stems or roots, plants can produce genetically identical copies of themselves. Leaves, stems and roots are normally involved in producing, storing and transporting food, and are known as vegetative organs. **Asexual reproduction** involving these organs is therefore called **vegetative propagation.** In the wild, strawberry plants, onions, spider plants and potatoes often reproduce by vegetative propagation.

Farmers and flower growers also take advantage of the ability of some plants to reproduce asexually. There are several advantages to creating genetic replicas of known plants. For one, if you select a parent plant that has all the characteristics you desire, you can ensure the production of identical offspring. Plants grown from vegetative organs are frequently stronger than those grown from seed, since they have a "head start" in their growth. Some plants, such as naval oranges and seedless grapes *must* be propagated by vegetative means since they have no seeds and cannot reproduce sexually.

In this investigation you will attempt to **clone** one or more plants by means of vegetative propagation. The directions you will be given are rather general so that you may experiment with various techniques and types of plants.

PURPOSE:

- How can plants be cloned using vegetative propagation?

MATERIALS:

One or more plants, as listed in the procedure or as supplied by your teacher
Single–edged razor blade or scalpel
Growing medium: choice of soil, sand, vermiculite and water
Pots or other containers in which to grow your plants
Labels or marking pencil

PROCEDURE:

Various methods of vegetative propagation may be used to clone plants. The most common methods, and plants adapted to each, are listed below:

Stem Cuttings:

Pieces of stem, approximately 10 cm long, can be cut and planted in moist sand, vermiculite or even in plain water. These cuttings should include several leaves, so that the growing plant can continue to produce food through photosynthesis. New roots will eventually grow from the cut stem.

Stem Cutting

Recommended plants: Coleus, geranium, Tradescantia, ivy, impatiens.

Leaf Cuttings:

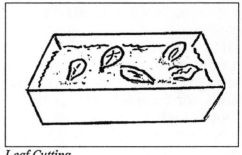

Leaf Cutting

If a leaf of certain plants is removed and placed on wet sand or vermiculite, new plants will grow from it. In certain cases, such as with Bryophyllum, entire new plants arise between the notches of the detached leaf. In other instances, such as with Begonia rex, shoots of new plants arise from the cut veins of the detached leaf.

Recommended plants: Bryophyllum, Begonia rex, African violet

Runners:

Certain plants produce modified, horizontal stems called runners. When the growing end of the stem touches the soil, it takes root and a new plant begins to grow.

Recommended plants: strawberries, spider plant, eel grass (aquatic)

Rhizomes:

Similar to runners, rhizomes are horizontal stems that grow below the ground, instead of above. New plants spring from the rooted tips of the rhizomes.

Recommended plants: grasses, lily of the valley

Tubers:

Fleshy underground stems, such as the white potato, contain buds, or "eyes" from which an entire new plant may grow. Cut a piece of potato containing a bud, and plant it on wet sand or vermiculite.

Recommended plants: white potato

Bulbs:

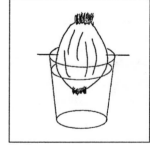

Bulb

A bulb is actually an enlarged underground stem that is wrapped in leaves. Many bulbs require rather specific temperature conditions in order to grow a new plant. The onion bulb, however, is not quite so particular. Suspend an onion in water and watch for the growth of roots and a new plant

Recommended plants: onion, garlic. Also try narcissus, hyacinth.

Fleshy Roots:

Underground, fleshy roots may give rise to new plants. A piece cut from such a root will give rise to several new roots, and subsequently several new plants. These may be suspended in water in the same manner as the bulbs.

Recommended plants: sweet potato, carrot, horseradish

Choosing from the plants available, select one or more methods, as directed by your teacher, and propagate a new plant. Label your experimental plants with the date and your initials. On your data sheet indicate the type of plant you are attempting to clone, as well as a description of the method of propagation you are using (include a sketch). Observe your experiment on a regular basis, looking carefully for signs of new plant growth. Prepare a data table and record your observations, such as the development of roots or the growth of shoots. Continue making observations until your new plant resembles its parent, or until directed to stop by your teacher. If you are conducting more than one propagation experiment, be sure to prepare separate data tables for each.

DATA:

Describe your experiment, including a sketch, data table, procedure and results.

CONCLUSION:

1. What are the vegetative parts of a plant, and why are they so named?

2. Explain why it may be an advantage for a farmer or horticulturalist to propagate plants by asexual means.

3. Is a plant that is grown from a stem cutting really a new individual? Explain.

4. Vegetative propagation is also known as regeneration. Is this an accurate term? Explain.

5. Explain why a field of potatoes grown from seeds will show more variation than a similar field grown from tubers.

SUGGESTIONS FOR FURTHER STUDY:

- Design and conduct an experiment to answer one or more of the following questions:

    Will leaf or stem cuttings grow better in wet sand or water?
    How does the length of a stem cutting affect its success?
    Does the presence of leaves on a stem cutting affect its success?
    Does the age of a Bryophyllum leaf affect its ability to regenerate?

    Have your experiment approved by your teacher before proceeding. Write a report of your experiment and its results.

- Research the methods by which fruit growers propagate bananas. Report on your findings.

- Design and conduct one or more experiments to determine the effect of plant growth hormones on vegetative propagation. Commercial preparations of hormones are available in nurseries and from garden supply dealers. Obtain approval from your teacher before beginning your experiment. Report on your results.

REGENERATION IN PLANARIA: TWO HEADS ARE BETTER THAN ONE

INTRODUCTION: Most plants, and many animals have the ability to replace lost body parts. This ability, known as **regeneration**, involves the same types of processes seen in a developing embryo – cell division, growth and **differentiation**. Regeneration usually involves only the replacement of a missing part, such as the growth of a new branch from the stem of a tree that has been pruned, or the growth of a leg that has been lost by a salamander. In some organisms, regeneration and **reproduction** may overlap. For example, if one of the rays (arms) of a starfish is removed, not only will the remaining animal regenerate a new ray, but the removed ray, providing it contains at least some of the central portion of the animal, can grow into an entire new starfish! Higher vertebrates have lost most of their regenerative capabilities, but still can manage to heal wounds, such as the replacement of skin and bone, as long as the damage is not too extensive. In this investigation, you will examine the regenerative capabilities of a planarian flatworm.

PURPOSE:

- How is regeneration accomplished in the planarian?

MATERIALS:

3–4 planaria	Cotton swabs
Petri dishes	Single–edged razor blade
Dissecting microscope or	Pond or aquarium water
hand lens	Glass microscope slide

PROCEDURE:

In this investigation you will perform 1–4 operations on planaria, according to directions from your teacher. In each case you will cut the planarian into one or more pieces, separate the pieces into labeled Petri dishes, and observe the pieces for healing and regeneration over a period of two weeks. Planaria may be transferred between dishes using a cotton swab. When excited, they tend to curl–up into a tight ball. Therefore, before you can cut them, or make any observations, you must wait for the planarian to relax and elongate. When making your cuts, place a planarian in a drop of pond or aquarium water in the center of a glass microscope slide (See illustration 1). A dissecting microscope, or hand lens can be helpful if your specimens are small. After your cuts are made, place each piece of cut planarian into a separate Petri dish, half-filled with pond or aquarium water. Label each dish precisely with the type of cut made, the section of planarian the piece represents, and the date. Cover the dishes, and store them in a dark area at room temperature (20–

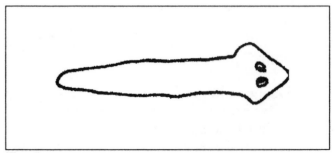

illus. 1

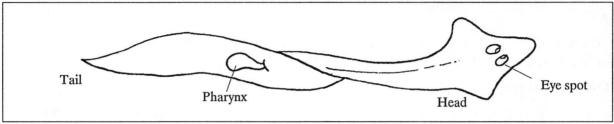

illus. 2

25°C). If any of your cut pieces die during the experiment, throw them away. Replace any water that evaporates during the observation time. Do not feed your planaria during regeneration.

The body of the planarian is rather simple. It is divided into three basic parts, a head region, the pharynx region, and a tail region. (See illustration 2). For your first operation, you will cut, horizontally, across the planarian in the area of the pharynx, so that you have two equal pieces. (This is essentially the manner in which a planarian reproduces asexually). Place each piece into an appropriately labeled dish, and keep in a cool dark place. Examine each piece every day or two. A hand lens or dissecting microscope can help to see the details of new growth. Make sketches on your data sheet to show the appearance of each piece at the start of the experiment, and then on a periodic basis as observable changes occur. Observe when the wounds heal, and note this on the data sheet. Observe the appearance of regeneration buds. Describe these, and note the date of their appearance on your data sheet. Continue making observations until the pieces either die or regenerate completely.

Your teacher will tell you how many more operations you are to perform. For each of your other operations, first write a question that you would like to answer regarding the regenerative abilities of planaria. For example, some questions that may be investigated include:

"Can smaller pieces regenerate into entire planaria?"
"Can a longitudinal cut result in two new planaria?"
"Will an oblique cut (one made on an angle) result in normal regeneration?"
"Can a partial cut result in two heads or two tails?"

Think up your own question to answer, or use one of those listed here. Illustration 3 shows possible cuts that you can make. There are many others as well, so don't let these suggestions limit your creativity. Write your question, and a sketch of the cuts you will make to answer it, on the data sheet. With teacher approval, conduct your experiment.

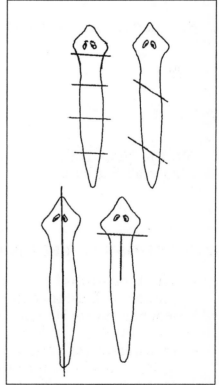

illus. 3

DATA:

Draw sketches and make observations on a separate piece of plain white paper. Use a separate piece of paper for each planarian you cut.

CONCLUSION:

1. How do regeneration buds differ from the rest of the piece on which they appear?

2. How does the new growth differ in appearance from the original piece of planarian?

3. According to your data from the first cut, how long does a piece of planarian take to regenerate a new body? Are your other experiments consistent with this timing?

4. Do you notice any differences between regeneration from the head (anterior) piece and the tail (posterior) piece? Explain.

5. Summarize the information you learned from your individual experiment.

6. Do the planaria that result from regeneration appear just like "normal" planaria? Explain.

SUGGESTIONS FOR FURTHER STUDY:

- Think of still other questions that can be investigated using regeneration in planaria. Design experiments to answer each of your questions. With teacher approval, conduct these experiments and report on your findings.

- Investigate regeneration in tadpoles or starfish. Design an experiment and obtain teacher approval. Conduct your experiment and report on your results.

WHAT IS THE FUNCTION OF A FLOWER?

INTRODUCTION: Angiosperms are plants that produce flowers. There are hundreds of thousands of different types of these plants, but they all have at least one thing in common; their organ of sexual reproduction is the flower. Flowers can vary dramatically from one another. Some, like roses and marigolds, have showy, fragrant **petals.** Flowers of the oak tree, on the other hand, have no petals at all. Since some flowers seem to function quite nicely without petals, these organs obviously are not essential to the functioning of all flowers. Organs such as these are know as **accessory organs.** The **essential organs** of a flower are those that produce the male or female gametes, or sex cells. When a male sex cell, enclosed in a **pollen grain,** fertilizes a female cell, or **ovule,** a **seed** is produced in which an embryo plant is found. Some flowers may have only the pollen-producing male organs. This male organ is called a **stamen.** Other flowers contain only the female organ, or **pistil.** There are also many flowers that contain both.

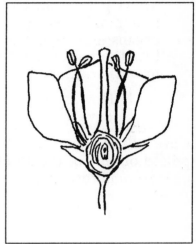

illus. 1

Even though flowers may vary significantly from one another, they all must be structurally adapted for sexual reproduction. In this activity you will examine the various parts of a **complete flower,** one containing petals, sepals, stamens and pistil, and relate each of these structures to its reproductive function. Illustration 1 shows the parts of a "typical" flower. Refer to this illustration when examining your specimen.

PURPOSE:

- How is a flower adapted for sexual reproduction?

MATERIALS:

Gladiolus, or other flower
Scalpel or single-edged razor blade
Transparent tape
Glass slide and coverslip
Stereomicroscope or hand lens
Compound microscope

PROCEDURE:

A. Examine the **accessory organs** of your flower. These are the parts that are not absolutely necessary for reproduction. The outermost whorl, or circle of specialized leaves, contains the **sepals.** In some flowers the sepals are green, in others they are brightly colored. Carefully remove the sepals from your flower and tape them onto a blank piece of paper so they can be clearly seen. Label the sepals.

    1. What might be the function of the sepals of a flower?

The remaining whorl of specialized leaves are the petals. Petals may be arranged in many different ways.

　　2. Describe the arrangement of the petals of your flower. How many petals are there?

Carefully remove the petals and tape them on the sheet of paper next to the sepals. Label the petals.

　　3. Suggest a function for the petals.

B. Examine the stalk–like organs, the **stamens**, which are also arranged in a whorl pattern in the center of the flower blossom. These are the male reproductive organs and consist of two parts, the stem–like **filament** supporting the enlarged top, or **anther**. (See illustration 2).

　　1. How many stamens does your flower have?

Carefully remove all the stamens. Look for pollen on one of the **anthers** using a hand lens or stereomicroscope.

　　2. What does the pollen look like?

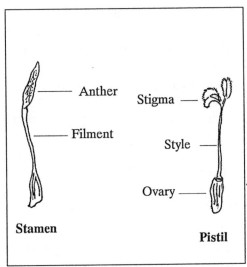

illus. 2

If pollen is present on the anther, gently shake it over a glass microscope slide to which a drop of water has been added. Add a cover slip and observe the pollen grains under a compound microscope.

　　3. On your data sheet, sketch a few pollen grains in detail. Tape the stamens to your paper, below the petals and sepals, and identify them. Label the anther and filament on one of the stamens.

C. The remaining, central portion of the flower is the **pistil**, the female reproductive organ. The pistil is composed of an enlarged, oval **ovary** at the base, a slender stalk–like **style**, and a top portion called the **stigma**. (See illustration 2) Examine the pistil of your flower. Using a hand lens, look at the stigma. Touch it.

　　1. Describe the appearance and surface of the stigma.

Remove the pistil from its base, the receptacle, but cut it with a scalpel below the ovary. **Be careful when cutting with a scalpel or razor blade. These are sharp instruments.** Now cut the ovary in half, crosswise. (See illustration 3). Take the upper half of the ovary, with the style and stigma attached, and tape it to your paper with the other flower parts. Identify the pistil. Label the ovary, stigma and style.

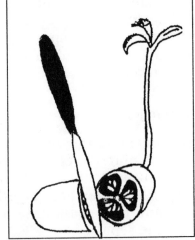

illus. 3

Look at the cut portion of the ovary through a stereomicroscope or hand lens. Notice that the ovary is divided into a number of separate compartments. Within each division are the small, white **ovules**, each containing an egg.

2. Draw a cross section of your ovary, showing the correct number of chambers and the placement of the ovules. Label your sketch appropriately.

DATA:

A. 1._____

2. _____

3. _____

B. 1._____

2. _____

3. SKETCH OF POLLEN GRAINS:

C.1 _____.

2. SKETCH OF OVARY X–SECTION:

CONCLUSIONS:

1. What is the reproductive function of each of the following flower parts?

    a) anther:

    b) stigma:

    c) ovary:

2. Why are all the other parts of the flower considered to be "accessory" organs, instead of "essential" ones?

3. What is meant by pollination? Fertilization?

4. Since pollen (containing sperm) is produced in the anther of a flower, and the egg is located inside the ovary of the same, or another flower, suggest a method by which the sperm could reach the egg so that fertilization can occur.

5. How is the structure of the stigma adapted to aid pollination?

6. Suppose you found two flowers growing next to each other in a field. One had bright orange petals with a strong sweet fragrance. The other had no petals at all. Describe the most likely method by which each is normally pollinated.

7. Monocot and dicot plants are identified by the number of flower parts they have. Flowers whose parts are in multiples of 4 or 5 are dicots, while monocots usually have flower parts in multiples of 3. Review your data and determine whether your flower is a monocot or dicot.

SUGGESTIONS FOR FURTHER STUDY:

- When pollen grains land on the stigma of a flower, they begin to grow a pollen tube. This tube grows down through the style until it reaches the egg within the ovule. Fertilization may then occur when a sperm nucleus travels down this tube and fuses with an egg nucleus. Pollen tube growth can be observed in the laboratory by supplying the pollen grains with a sugar solution whose concentration closely approximates that of the natural plant's. Sucrose or glucose solutions between 2% and 25% may be appropriate. Flowers must be ripe enough so that a gentle tap releases pollen from the anthers. Pollen grains should be added to clean Petri dishes containing the desired sugar solutions and incubated at 27°C. Some pollen tubes will begin germinating in less than 30 minutes, while others may take one or two days. Frequent observations, using a hand lens or stereomicroscope will be required in order to collect accurate data.

Plan and carry out one or more experiments to determine:

A. The rate of growth of the pollen tube for one species of plant (vary sugar concentrations.)

B. The percentage of germination of various types of pollen (at a given sugar concentration.)

C. The effects of temperature, pH, light, etc., on the rate of pollen tube growth.

Other experiments may be attempted with permission from your teacher. Report your experimental results.

WHAT ARE THE CONDITIONS NECESSARY FOR SEED GERMINATION?

INTRODUCTION: Plants that reproduce sexually produce seeds. Once the seeds have matured and have been distributed, they must encounter the proper conditions for growth, or **germination**. Most seeds require some moisture. You are probably aware of the need to water growing plants. But how much water do various seeds need? What other factors might be important for the germinating seed? Is oxygen required? What about soil? What light conditions, if any, are required? How about a specific range of temperatures? Do these conditions vary with the type of seed germinating? In this investigation, you will explore some of these questions.

PURPOSE:

- What conditions are required for seed germination?

MATERIALS:

Seeds (select from the choices available in your classroom). Ten seeds of each type will be needed for each variable you are testing.

Containers for germinating seeds: See your teacher for the materials available. Peat pots, small flower pots, empty milk cartons, or petri dishes are some appropriate containers.

PROCEDURE:

1. To begin, select one **variable** that you think is important for seed germination. On your data sheet, write an **hypothesis** relating to your variable and its effect on seed germination. Possible variables include, but are not limited to, light, the amount of moisture, soil, temperature and depth of planting.

2. Design a simple experiment to test your hypothesis. Include a **control** in your experiment. Use at least 5 seeds for each of your set–ups. On a separate sheet of paper, describe your **experimental design**, including a sketch, if appropriate.

3. Decide what **data** you should collect, and how often it should be recorded. Construct an appropriate data table on the sheet with your experimetal design.

4. Obtain the approval of your teacher, then proceed with your experiment. Record data until you have conclusive results.

5. At the conclusion of your experiment, review your data and form a **conclusion** relating to your hypothesis. Be sure your conclusion is consistent with your data. Write your conclusion in the appropriate space on your data sheet.

6. If time and materials permit, repeat the entire procedure with a different variable.

7. Share your conclusions with the class, as directed by your teacher.

DATA: Variable being tested _____

Hypothesis _____

_____

Conclusion: _____

_____

CONCLUSIONS:

1. Did you support or negate your hypothesis? Explain.

2. According to the class data, what conditions are necessary for seed germination?

SUGGESTIONS FOR FURTHER STUDY:

*   You based your conclusion on data collected using only one type of seed. Will your conclusion hold true for other types of seeds? Repeat your initial experiment using at least three different seed types. Is your conclusion still valid?

*   Some seeds require rather extreme conditions in order to germinate. For example, the seeds of certain pine trees will not germinate until they have been in a forest fire! This requirement is apparently an evolutionary adaptation in these particular plants. Research this, or any other plant with unusual requirements for seed germination. Describe the plant's requirements, and try to determine of what value this condition could be to the survival of that plant species.

HOW DOES A SEED GROW?

INTRODUCTION: Seeds develop from the fertilized ovule of the flower. Contained within the seed is the tiny plant **embryo**. This embryo is nourished and protected by the seed leaves, or **cotyledons**. Some plants, such as corn and grass, contain a single cotyledon and are therefore known as **monocots**. Common trees, most flowers and vegetables have seeds with two cotyledons, and are known as **dicots**. The embryo itself is seen to be composed to two distinct parts, the embryonic root, or **hypocotyl**, and the embryonic shoot, or **epicotyl**. At the tip of the epicotyl are small leaves, known as the **plumule**. (See illustration 1).

If a seed is given the proper conditions (see investigation B30), it will usually sprout, or **germinate**. Rapid cell division (mitosis), differentiation and growth proceed to transform the seed into a new, mature plant. Growth in higher plants is restricted, however, to specific areas called **meristems**. At the tips of the roots and stems are found the **apical** meristems, regions that are responsible for growth in length. Many plants also contain a lateral meristem, or **cambium**, that is responsible for growth in the diameter of roots and stems. (See illustration 2).

In this investigation you will observe the growth of various embryonic structures from a seed into a small plant. You will also measure the rate of growth of these various structures so that the role of meristem tissue may be observed.

PURPOSE:

- How does a seed grow into a plant?

MATERIALS:

Soaked lima or kidney beans	Hand lens
Two 250–400 ml beakers	Compound microscope
Cotton or vermiculite	Single–edged razor blade
Glass marking pencil	Blotting paper or paper towels
Waterproof marking pen	Sodium hypochlorite (dilute solution)
Metric ruler	Goggles
Microscope slide and cover slip	

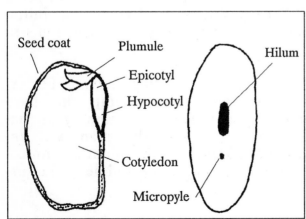

illus. 1: Bean seed

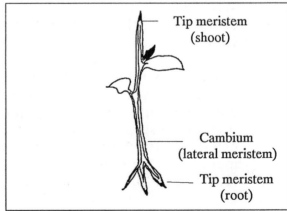

illus. 2: Meristems

PROCEDURE:

A. Structure of Seed:

Examine the external structure of the bean seed as seen in illustration 1. Using a hand lens, look along the vertical edge and find the scar that served as the attachment of the ovule to the ovary. This scar is called the **hilum**. Just below the hilum is a tiny hole, the **micropyle**, that marks the spot where the pollen tube entered the ovule.

Carefully remove the **seed coat** from the soaked bean seed. Separate the two cotyledons to expose the embryo. Observe the embryonic structures with a hand lens. Look for the hypocotyl, epicotyl and plumule.

On your data sheet, make one or more sketches of the seed and embryo, labeling all the structures you can see.

B. Seed Preparation and Planting

1. Place about 10 soaked bean seeds in a dilute solution of sodium hypochlorite for 15 minutes. (This chemical will prevent the formation of mold on your seeds.) Wear goggles during this procedure. Remove the seeds from the solution and rinse them in cold water. Set them aside on a paper towel.

2. Prepare two beakers by lining each with blotting paper or paper towels. Fill the inside of the beaker with cotton or vermiculite. Add water to the beaker so that the cotton or vermiculite is moist, but not soaking wet. There should be no more than two centimeters of water in the bottom of each beaker. (See illustration 3). Write your name and date on each beaker.

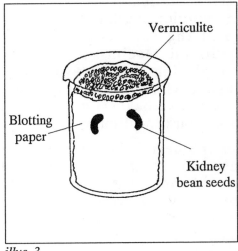

illus. 3

3. Without tearing the blotting paper, insert four bean seeds, scattered around the beaker, between the glass and the paper.

4. The other beaker will contain four embryos, without the cotyledons attached. Prepare these seeds as follows:

   a) With a single–edged razor blade, cut each of four seeds in half, horizontally.

   b) Discard the lower half (with the micropyle) of each seed.

   c) Carefully open the remaining halves to expose the embryo. Remove and discard the piece of cotyledon that does not support the embryo plant. (Note: If the embryo breaks during this procedure, discard the seed and select another).

d) In the second beaker, plant each of the four embryos as you did the complete seeds. Label this beaker "embryos only."

C. Growth and Development of Seeds:

Set the beakers in a warm, dark location. Keep the blotting paper (and seeds) moist by replacing any water lost to evaporation. If any seeds seem to be rotting, remove them at once. Observe the seeds daily and look for the first signs of germination. Seeds should begin to germinate within 3 days.

At the first sign of germination, record the date and describe the growth that has occurred. Continue these observations daily (or as directed by your teacher) for the next two weeks. Include sketches in your descriptions. Record your data on Data Table 1.

On the fifth day of growth (or when the hypocotyl is about 3 cm long), carefully remove two similar seeds from each beaker, after noting their original positions. Place the seedlings on paper towels and gently blot them dry. Measure the length of each hypocotyl, from its tip to the spot where it meets the cotyledon.

Record the length of each root on Data Table 2. Using a waterproof pen and a metric ruler, mark each root at 3 mm intervals, with the first mark as close to the tip of the root as possible. (See illustration 4). Return the seedlings to their proper beakers, carefully replacing them between the blotting paper and glass, as close to their original locations as possible. Be sure the beakers are moist, and return them to a warm, dark location.

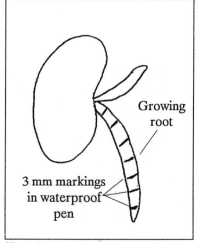

Growing root

3 mm markings in waterproof pen

illus. 4

For the next 4 days, remove the marked seeds from each beaker and measure the length of the roots. Record your measurements in Table 2. Also measure the distance *between* each of the markings you made on the roots. At the end of your observation period, compute the *average* rate of growth for the seeds with and without cotyledons.

On the fifth day of growth, also remove one of the *unmarked* seedlings. With a single–edged razor, cut off approximately 3 mm from the tip of the root. Place the root tip in a drop of water on a glass microscope slide. Cover the root tip with a cover slip (press down firmly to flatten the root tip somewhat) and observe under the low power of your microscope. Look for the presence of small root hairs. Sketch these on your data sheet.

DATA: A. SKETCH OF A BEAN SEED

DATA TABLE 1: OBSERVATIONS ON THE GROWTH OF BEAN SEEDLINGS

Date	Whole Seeds		Embryos Only	
	Sketch	Description	Sketch	Description

DATA TABLE 2: GROWTH OF ROOTS

		Day									Average rate of growth
		1		2		3		4		5	
	Length	Distance between marks	Length	Distance between marks	Length	Distance between marks	Length	Distance between marks	Length	Distance between marks	
Whole Seeds	1	3mm									
	2	3mm									
Embryo Only	1	3mm					.				
	2	3mm									

C. Sketch of Root Tip Showing Root Hairs

CONCLUSIONS:

1. Which part of the hypocotyl seemed to grow the fastest?

2. Was there a difference in the rate of growth of the hypocotyl from complete seeds and partial cotyledons? If so, describe the differences.

3. Were there any other observable differences between the growth of seedlings from complete seeds and the growth of seedlings from partial cotyledons? Describe these.

4. How can you account for any differences noted in response to questions 2 and 3?

5. Which is the first plant organ to grow from a germinating seed? How might this be a survival adaptation for the plant?

6. From which embryonic structure does the plant's stem arise?

7. Approximately how long does it take a seedling to develop leaves capable of photosynthesis? What does the seedling use for food until then?

8. What eventually becomes of the cotyledons during seedling germination?

9. What could be a function of the root hairs that you observed under the microscope?

SUGGESTIONS FOR FURTHER STUDY:

- Repeat this investigation using a monocot, such as a corn seed. Describe any differences in the growth pattern of the two types of seedlings.

- Repeat this investigation, but include measurements of the length/diameter of the stem, in a manner similar to that done with the root. Describe the similarities and differences in the growth of the stem and root.

- Investigation the effects of auxins on the germination and growth of seeds.

- Carry out an experiment to determine what percentage of any specific variety of seeds will germinate under normal conditions.

A MODEL OF DNA

INTRODUCTION: In the 1950's, James Watson and Francis Crick were credited with discovering the structure of the **DNA** molecule. Working primarily from data obtained by X–ray crystallography techniques, they developed a model shaped like a double helix. This model helped usher in a whole new field of biology, often called molecular genetics, which in turn has led to areas as significant as genetic engineering and gene therapy.

The fundamental structure of the DNA **double helix** is based on a unit called a **nucleotide** (see illustration 1). This unit, composed of a sugar (deoxyribose), a phosphate group, and a nitrogenous base (purine or pyrimidine), is formed into an extremely long chain, or **polymer**. In DNA, two such chains are bonded together and twisted, resulting in a double helix structure. Illustration 2 is a representational diagram of a small section of a DNA molecule.

In order to more accurately visualize the three–dimensional structure of DNA, you will construct your own model in this investigation.

PURPOSE:

- How can you construct a model of DNA?

MATERIALS:

Cheerios
White straws
Straws of 4 different colors
Ring stand with 2 ring clamps
1 meter of string
Toothpicks

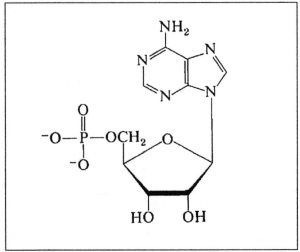

illus. 1 Structural Formula of an Adrenine Nucleotide

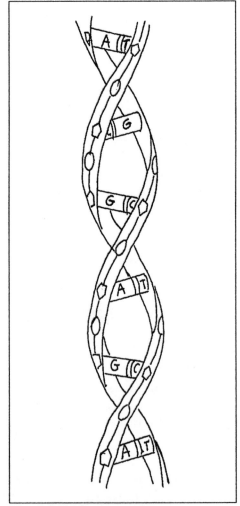

illus. 2 DNA Representation

PROCEDURE:

A. PREPARATION OF MATERIALS

Phosphate molecules:

These will be represented by pieces of **white straws**. Cut several white straws into 2–cm lengths.

Sugar molecules:

These will be represented by individual Cheerios. Use as is.

Nitrogenous Bases: adenine, guanine, cytosine and thymine

These will be represented by the four different colored straws. Cut two or three of each color into 3–cm lengths. Assign each base its own color. Record the color of the straw representing each base on your data sheet.

B. SET UP (see illustration 3).

Attach the ring clamps to the top and bottom of the ring stand. Cut the 1 meter length of string in half. Tie one end of each piece of string to the bottom ring clamp, approximately 6 cm apart.

Onto one of the strings, slide a white straw (phosphate), then a Cheerio (sugar). Continue stringing these two "molecules" until you have reached the top of the string.

Tie the upper end of the string to the top ring clamp so that the string is in a taut, vertical position.

Repeat the procedure with the other string.

Put two different colored straws onto a toothpick. Consult your code and remember that only adenine can pair with thymine, and only cytosine can pair with guanine.

Stick the toothpick into two opposite cheerios.

Continue adding "base pairs" until you have completed your molecule. Loosen the top ring clamp, and rotate it in a clockwise direction to produce a helix. Retighten the clamp.

CONGRATULATIONS! Your DNA Model is Complete!

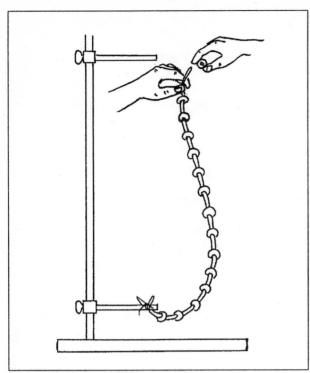 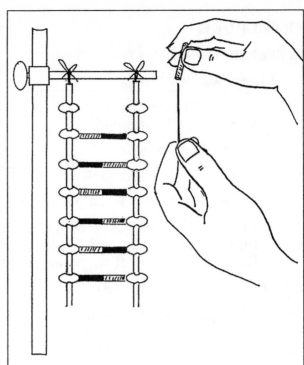

illus. 3 Preparation of DNA Model

DATA:

Color code for nitrogen bases:

Adenine = _____

Thymine = _____

Cytosine = _____

Guanine = _____

CONCLUSIONS:

1. In your model, what substances represent one nucleotide?

2. How many nucleotides are in one strand of your model? Does the other strand contain the same number? Explain.

3. How many of each TYPE of nucleotide (A,T,C or G) are in **EACH** of your DNA strands? What do you notice about these numbers?

SUGGESTIONS FOR FURTHER STUDY:

- Using a molecular model kit, construct a more accurate model of DNA, showing the positional relationships among the actual molecules.

- Read the book *The Double Helix: A Personal Account of the Discovery of the Structure of DNA*, by James Watson (Antheneum, 1963). Write a report on this work. Include a summary as well as your impressions of the man and his relationship with Frances Crick.

- Investigate the work of Rosalind Franklin. How did her research help Watson and Crick? Do you believe she was "overlooked" when credit for the discovery of DNA was given? Write a short paper on this subject.

- Discuss the technique of X–ray crystallography. How is it done? What does it show? Is it still a valuable research tool? Prepare a short paper on your findings.

CONSTRUCTING AND INTERPRETING A HUMAN KARYOTYPE

INTRODUCTION: Approximately 30 years ago, J.H. Tijo and Theodore Puck, at the University of Colorado, first showed that the normal chromosome number in humans is 46. Since then, rapid scientific advances have been made in the analysis of these **chromosomes**. One method of analysis, called a **karyotype**, involves the pairing and sequencing, from actual photographs, of chromosomes from an individual's cells. The technique begins when a cell, such as a white blood cell, is "frozen" during the metaphase stage of mitosis. In this stage, all the chromosomes are double stranded, attached by a **centromere**, and visible under a microscope. A photograph of the cell is taken through the microscope, enlarged, and is then ready for analysis. (See illustration 1). Each chromosome is cut out and matched with its **homologous pair**. The matching is done according to the length of the chromosome and the location of the centromere. Once the **autosomal** pairs are matched, they are arranged in size order, according to a standard pattern developed in 1960. (See illustration 2). The last, unmatched pair of chromosomes, the **sex chromosomes**, are kept separated at the end of the karyotype. In normal females, there will be two X chromosomes; in normal males, one X and one, shorter Y.

The primary use of karyotyping is in the identification of chromosomal aberrations as they relate to **congenital** genetic disease. An unborn fetus may have its karyotype completed before it is 16 weeks old. A sample of the fetal cells may be collected through **amniocentesis**, and analyzed if a possible genetic defect is suspected. A newer technique for obtaining fetal cells, called chorionic villi sampling (CVS), is currently being developed in order to be able to perform fetal chromosomal analysis at a much earlier date.

PURPOSE:

- To prepare and interpret human karyotypes.

MATERIALS:

Copies of karyotype plates 1 + 2
2 copies of karyotype analysis sheet
Scissors
Transparent tape or glue

PROCEDURE:

1. Working with one metaphase plate at a time, cut out each of the chromosomes. To make this job easier, cut each chromosome with a small, rectangular border around it. The sex chromosomes in each plate are labeled to make identification easier.

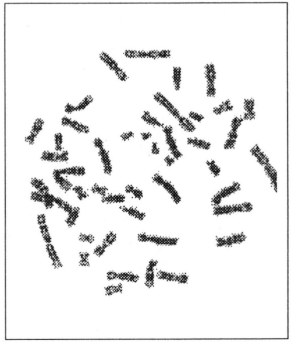

illus. 1 Human Chromosomes

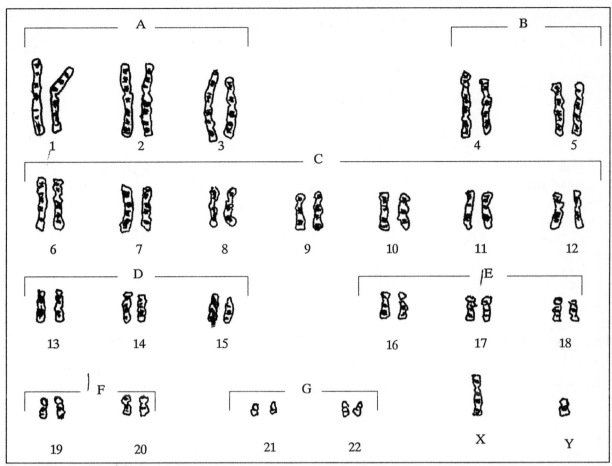

illus. 2 Sequencing of Chromosomes

2. Using illustration 2 as a guide, arrange each of the autosomal chromosomes in pairs, paying attention to the location of the centromere and length of the "arms." Arrange the pairs in descending order of length and position of the centromere.

3. Paste or tape the chromosome pairs on a karyotype analysis sheet. Be sure to identify each analysis sheet with the karyotype plate number.

4. Refer to your karyotype and complete the bottom portion of the analysis sheet. The following definition will be needed:

   trisomy: three chromosomes in a set, instead of the normal two.

CONCLUSIONS:

1. Summarize the genetic conditions you found in the two karyotypes you constructed.

2. Describe the events that might have occurred during meiosis to account for the chromosomal abnormality you found.

3. How does an autosome differ from a sex chromosome?

4. In your own words, of what value is karyotyping, and genetic screening in general?

5. What drawbacks might there be to the processes mentioned above?

SUGGESTIONS FOR FURTHER STUDY:

*   Using current references, investigate the physiological consequences of the chromosomal abnormality you found in the karyotype from this investigation.

*   Contact a local hospital and determine if they have a genetic counseling department. If they do, arrange to visit their facilities. If possible, interview one of the counselors. Prepare a report on the services offered.

*   Prepare a report on the pros or cons of genetic screening.

*   Individuals on trial for criminal acts have, on occasion, tried to use their genetic abnormality as a reason for their actions. They have pled not guilty due to the effects of their genetic problem. Research the relationship between genetic disease and the law, and report your findings

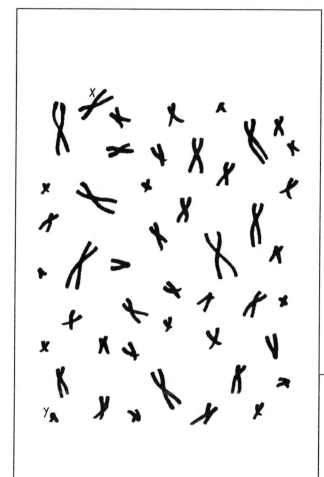

Plate #1

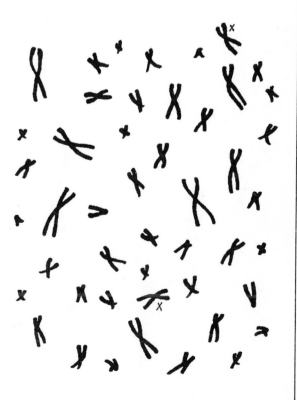

Plate #2

KARYOTYPE ANALYSIS SHEET

Plate # _____

Analysts Name _____

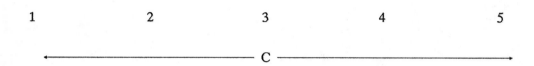

←————————— A ————→   ←———— B ————→

1   2   3   4   5

←——————————— C ———————————→

6   7   8   9   10   11   12

←———— D ————→   ←———— E ————→

13   14   15   16   17   18

←—— F ——→   ←—— G ——→   Sex Chromosomes

19   20   21   22

Sex of Subject _____

Abnormality
Present _____

PREPARING A HUMAN PEDIGREE CHART

INTRODUCTION: Show dogs and race horses are not the only animals that have a pedigree. You can have one too, since a **pedigree** is nothing more than a family tree that traces the inheritance of a particular genetic trait. A pedigree chart is a visual representation that shows the **phenotypes** of related individuals, and provides a basis for attempting to determine their **genotypes**. The pedigree chart below represents three generations of a family. Each symbol represents a specific individual, as explained in the key. In this example, a male and female married and had four children, two boys and two girls. One of these children got married and they, in turn, had two children, one boy and one girl. The individuals whose symbols are darkened show the trait that is being investigated, while the clear symbols represent individuals who do not show this trait.

SAMPLE PEDIGREE CHART

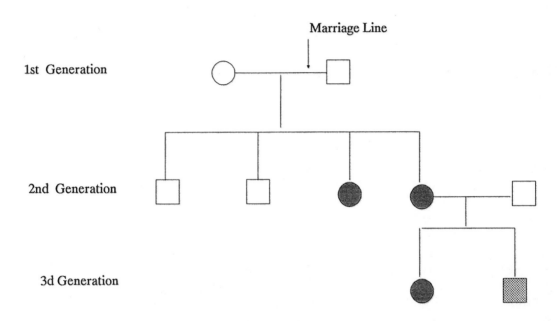

KEY TO PEDIGREE CHART

Symbol	Meaning
○	female without the trait
●	female with the trait
□	male without the trait
▦	male with the trait

Genotypes may be determined, in some cases, by analyzing the pattern of inheritance shown in the pedigree chart. In the example above, suppose the trait illustrated was a **recessive** trait inherited by a single pair of alleles. Let the **dominant** allele be represented by the letter **T**, and the recessive by **t**. In order for an individual to actually express a recessive trait, that person must be **homozygous recessive**, or have a **tt** genotype. An individual who does not possess the trait must have at least one dominant allele, and his/her genotype could be **TT** or **Tt**. By examining the pedigree chart it may be possible to determine the specific genotype of dominant individuals by looking at their offspring. In the example above, for instance, the first generation couple does not have the trait, but two of their children do. Those children must have the genotype **tt**. Their parents, therefore, must be **heterozygous**, or **Tt**. This is the only genotype possible so that they both show the dominant phenotype but still each pass on a recessive allele to their offspring.

In this investigation, you will examine some of your genetic traits and construct pedigree charts of your own.

PURPOSE:

- To construct pedigree charts of human genetic traits.

MATERIALS:

PTC paper

PROCEDURE:
Part A.

Most human traits are too complex to analyze with a pedigree chart. Some, however, like those described below, seem to be inherited by a single pair of alleles, and are either dominant or recessive. Examine yourself for each of the traits described. On Data Table 1, record your phenotype and genotype for each trait. Remember that you will know your complete genotype only if you possess the recessive phenotype. If you have the dominant trait you only know for certain that you possess at least one dominant allele (capital letter in your genotype). Since you can't be sure about your other allele at the present time, just write a ? in its place.

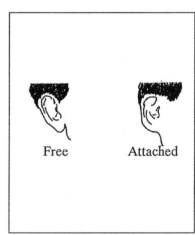

illus. 1 Earlobes

TRAITS:

Ear Lobes: Free earlobes are dominant (**L**) over attached earlobes (**l**). (See illustration 1)

Tongue Rolling: The ability to roll your tongue (into a "U" shape) when it is extended is dominant (**R**). The inability to perform this action is a recessive trait (**r**). (See illustration 2)

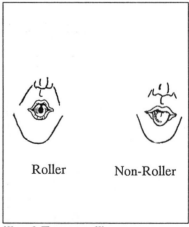

illus. 2 Tongue rolling

Eye Color: Although eye color is controlled by several genes, a single pair is responsible for blue eye color. Blue eyes are a recessive trait (**b**), while all non–blue eye color is dominant (**B**).

Hair Color: Red hair is a recessive trait (**h**), while all colors other than red are the result of a dominant allele (**H**).

PTC Tasting: The ability to taste a harmless chemical, abbreviated PTC, is a genetic trait. Tasting (**T**) is dominant over non–tasting (**t**). Get a piece of paper that has been soaked in PTC from your teacher. Put it in your mouth and chew it for a few seconds. You will notice a distinct bitter taste if you are a taster.

Widow's Peak: Pull back the hair from your forehead and look (in a mirror) at your hairline. If your hairline comes to a definite point in the center of your forehead, you have a widow's peak, and it's a dominant trait (**W**). A smooth, straight hair line is recessive (**w**).

Check the phenotypes of your parents for those traits in which you have incomplete genotypes. If either of your parents are homozygous recessive, you can fill–in your missing allele with a recessive one.

1. On your data sheet, explain why the above paragraph is valid.

2. For which traits do you have the dominant phenotype?

3. For which are you recessive?

Part B:

Prepare a pedigree chart for TWO of the traits you examined in Part A. In order to gather the information required for a pedigree chart, you must examine as many family members as possible. In selecting which traits to use for your charts, pick those in which family members show different phenotypes, if possible. If you are using the ability to taste PTC as one of your traits, be sure to bring home enough papers for your family members to taste. If your grandparents are available, include them in your charts. Draw your pedigree charts on a separate piece of paper and attach it to your lab report. When making your pedigrees, refer to the key shown earlier. Indicate the symbol that represents you by marking it with an asterisk (*). Beneath each individual on your pedigree charts write his or her genotype. Again, if one allele is unknown, use a ? in its place.

DATA: Part A

DATA TABLE 1: PHENOTYPES AND GENOTYPES OF TRAITS

Trait	Phenotype	Genotype
Shape of Ear Lobes		
Tongue Rolling		
Eye Color		
Hair Color		
PTC Tasting		
Widow's Peak		

1._____

_____

2._____

_____

3._____

_____

CONCLUSIONS:

Answer the conclusion questions below based on the following pedigree chart that shows the appearance of the tongue rolling trait in a family.

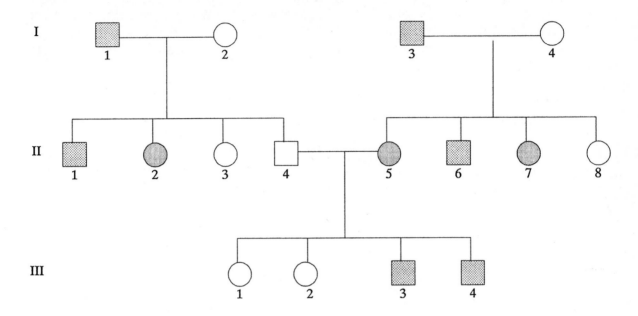

1. List the genotypes of each individual in the pedigree chart.

   First Generation: 1 _____   3 _____
                         2 _____   4 _____

   Second Generation: 1 _____   5 _____
                         2 _____   6 _____
                         3 _____   7 _____
                         4 _____   8 _____

   Third Generation: 1 _____   3 _____
                         2 _____   4 _____

2. If individuals 4 and 5 in generation II have another child, what is the probability that it will be a roller? Explain.

3. If individual 8 in generation II were to marry a man who is homozygous dominant for tongue rolling, what is the probability that their first child will be a non–roller? Explain.

SUGGESTIONS FOR FURTHER STUDY:

- Dr. Nancy Wexler, of the New York State Psychiatric Institute, and Dr. James Gusella, of Harvard University, recently conducted extensive studies on the epidemiology of Huntington's disease. The genetic patterns, shown in a specific population in which the disease was common, were analyzed on a massive pedigree chart. Their research led to the location of at least two genetic markers for this disease. Investigate this work, and summarize their studies, including any recent developments in the field.

THE BIOCHEMICAL EVIDENCE FOR EVOLUTION

INTRODUCTION: Ever since Charles Darwin's publication of *The Origin of Species,* scientists have been studying evolution by comparing anatomical structures of various organisms. During the past decade, however, advances in molecular genetics have opened a new field of evolutionary study. **Biochemical evolution** deals with the changes in the structure of **DNA**, and corresponding changes in the proteins of organisms. If the DNA and proteins of two organisms are quite similar, scientists believe that the organisms are closely related in an evolutionary sense. Conversely, if the DNA and proteins are dissimilar, the evolutionary relationship of the two organisms is a distant one. Biochemists have even deciphered a molecular clock, or rate of DNA mutation common to all organisms, by which they can calculate the approximate time at which various organisms diverged from a common ancestor.

In this activity, you will examine the DNA base sequence for a piece of the hemoglobin protein in humans, gorillas and horses. From this information you will determine the amino acid sequence for the protein fragment in each organism. Comparing the similarities or differences in this sequence will result in an understanding of the significance of biochemical evolution.

PURPOSE:

- How can biochemical studies aid in the understanding of evolutionary relationships?

MATERIALS:

None

PROCEDURE:

DNA carries the instructions for the synthesis of proteins in its sequence of nitrogen bases. The four bases, **adenine (A), thymine (T), cytosine (C) and guanine (G)** are arranged as a linear group of three to make up a single **codon.** This codon directs the synthesis of a complimentary strand of **messenger RNA (m–RNA).** Complimentary base pairing between DNA and RNA matches adenine with **uracil,** thymine with adenine, cytosine with guanine and guanine with cytosine.

The m–RNA codons then transfer their "message" to **transfer RNA (t–RNA),** which is responsible for carrying specific amino acids to a ribosome for incorporation into a protein. Scientists have determined which m–RNA codons correspond to which amino acids. This **genetic code** *seems* to be common to all organisms, and can be read just like you might translate the dots and dashes of the Morse code into letters and words. A copy of the genetic code is shown on the next page, in Table 1. The letters U, C, A and G on the left, top and right borders represent the m–RNA bases. The first base of a codon is represented by the letters on the left, the second base by the letters on the top, and the third by the letters on the right. To read the code, locate each of the letters corresponding to the m–RNA codon you would like to translate. At the junction of the three letters you will see a three–letter abbreviation standing for an amino acid. For example, the m–RNA codon that reads AGU would stand for the amino acid SER, which is an abbreviation for serine. A list of the names and abbreviations of amino acids found in the code is given in Table 2. Study the genetic code and answer the questions that follow.

1. What amino acid is coded for by the m–RNA codon GUA? AAG?

2. What is the DNA codon that corresponds to the m–RNA codons listed in question 1?

3. How many m–RNA codons can result in the amino acid glycine? What are they?

4. What does the m–RNA codon UAG result in? What do you think this means?

If you were given the DNA codons for a particular peptide, the first thing you must do is determine the complimentary m–RNA codons. Once you have these, the corresponding amino acids may be identified from the genetic code. In Table 3 on the data page you will find a 30–codon segment of the DNA responsible for a portion of hemoglobin, the protein that carries oxygen in red blood cells. This protein is found in many animals, including humans, gorillas and horses.

a. For each DNA codon listed, write the complimentary RNA codon.

b. Using the genetic code, identify the amino acid that corresponds to each codon. Record these in the appropriate space on Table 3.

Use the information from the table to answer the following questions:

5. How many DNA *bases* are different in the human and gorilla sequences?

6. How many *bases* are different between the human and horse?

Table 1: GENETIC CODE

| 1st Base | Second Base | | | | 3rd Base |
	U	C	A	G	
U	PHE	SER	TYR	CYS	U
	PHE	SER	TYR	CYS	C
	LEU	SER	Stop	Stop	A
	LEU	SER	Stop	TRP	G
C	LEU	PRO	HIS	ARG	U
	LEU	PRO	HIS	ARG	C
	LEU	PRO	GLN	ARG	A
	LEU	PRO	GLN	ARG	G
A	ILE	THR	ASN	SER	U
	ILE	THR	ASN	SER	C
	ILE	THR	LYS	ARG	A
	MET	THR	LYS	ARG	G
G	VAL	ALA	ASP	GLY	U
	VAL	ALA	ASP	GLY	C
	VAL	ALA	GLU	GLY	A
	VAL	ALA	GLU	GLY	G

Table 2: THE 20 AMINO ACIDS IN PROTIENS

Glycine	GLY	Lysine	LYS
Alanine	ALA	Arginine	ARG
Valine	VAL	Asparagine	ASN
Isoleucine	ILE	Glutamine	GLN
Leucine	LEU	Cysteine	CYS
Serine	SER	Methionine	MET
Threonine	THR	Tryptophan	TRP
Proline	PRO	Phenylalanine	PHE
Aspartic acid	ASP	Tyrosine	TYR
Glutamic acid	GLU	Histidine	HIS

7. Which two organisms are more closely related in terms of their DNA base sequence for this section of the hemoglobin molecule?

8. How many *amino acids* differ in the human and gorilla hemoglobin segment?

9. How many *amino acids* differ between human and horse?

10. Which two organisms are more closely related based on the sequence of their amino acids in the hemoglobin protein?

DATA:

1._____

2._____

3._____

4._____

5._____

6._____

7._____

8._____

9._____

10._____

Table 3: DNA CODONS FOR A 30–CODON SEGMENT OF THE HEMOGLOBIN GENE IN THE HUMAN, GORILLA AND HORSE

#	ORGANISM	DNA CODON	M–RNA CODON	AMINO ACID	#	DNA CODON	M–RNA CODON	AMINO ACID
1	human	GCA				CCA		
	gorilla	GAG				CCA		
	horse	GCA				CCT		
	human	GAG				TAC		
	gorilla	GAG				TAC		
	horse	GAG				TAC		
	human	GAA				CTT		
	gorilla	GAA				CTT		
	horse	GAA				CTA		
	human	CCG				AAA		
	gorilla	CCG				AAA		
	horse	CCG				AAA		
5	human	CTA			20	TGG		
	gorilla	CTA				TGG		
	horse	CTA				TGC		
	human	CAA				GGG		
	gorilla	CAA				GGG		
	horse	CAA				GGG		
	human	GAG				GGC		
	gorilla	GAG				GGC		
	horse	GAG				CTT		
	human	CAT				CAC		
	gorilla	CAT				CAC		
	horse	CGG				GAC		
	human	ACG				CTT		
	gorilla	ACG				CTT		
	horse	GAG				CTT		
10	human	CAA				CGG		
	gorilla	CAA			25	CGG		
	horse	CAA				CGA		
	human	GAG				CGG		
	gorilla	GAA				CGG		
	horse	GAA				AGT		
	human	CGG				ATA		
	gorilla	CGC				ATA		
	horse	CGC				ATA		
	human	GTA				CTC		
	gorilla	GTA				CTC		
	horse	GCA				CTC		
	human	GTA				TTT		
	gorilla	GTA				TTT		
	horse	GTA				TTT		
15	human	AAA				CAA		
	gorilla	AAA			30	CAA		
	horse	AAA				CAT		

CONCLUSIONS:

1. Examine the genetic code. How many amino acids have more than one codon? What might be the purpose of this redundancy?

2. Review your answers to procedure questions 5, 6, 8 and 9. How do you account for the discrepancy between the number of different *bases* and the number of different *amino acids* among the human, gorilla and horse?

3. What mechanism may be responsible for the changes in the base sequences of DNA from the horse to the gorilla to the human?

4. The horseshoe crab was recently reclassified into the same class as spiders (Arachnids). On what might this new classification be based?

5. Give reasons to support or reject the following statement: "Living organisms with similar proteins have a stronger evolutionary relationship than organisms with dissimilar proteins."

SUGGESTIONS FOR FURTHER STUDY:

*   Read the article "The Molecular Basis of Evolution," by Allan Wilson, in *Scientific American,* October 1985, pg. 164–173. Prepare a brief summary of the main points in the article.

THE SCOPE'S MONKEY TRIAL

INTRODUCTION: The theory of evolution, as developed by Charles Darwin and many other scientists, has, at its core, the idea that living organisms have changed over billions of years. Through the process of natural selection and survival of the fittest, more complex organisms have evolved from simpler ones over long periods of time. According to the scientific evidence compiled, humans did not appear on Earth until the relatively recent geological past.

Historically, the theory of evolution has met with resistance from various facets of our society, specifically certain religious groups. In fact, several states had laws prohibiting the teaching of this concept in public schools because they felt it contradicted the teachings of the Bible. In Tennessee, such a law was still in effect as recently as 1967. Back in the 1920's, however, a teacher named John T. Scopes decided to challenge this law, allowing himself to be "caught" teaching evolution in a public high school.

The Scope's trial attracted worldwide attention, and is one of the most famous legal cases in United States history. In this investigation, you will become more familiar with this case.

PURPOSE:

- What can be learned from a study of the Scope's Trial?

MATERIALS:

Library resources

PROCEDURE:

This is a research activity, and does not involve laboratory work. Using a variety of resources, investigate the Scope's trial. Information can be obtained not only from the library, but also your school's social studies department. If you have not read *Inherit the Wind*, you may want to do so at this time, but keep in mind that it is a play, and not a factual documentary.

During your research, answer the following questions about the case. Be sure your answers are in your own words. Use complete sentences and proper grammar. Attach a bibliography of your sources, written in the correct style.

1. Who were the two famous men involved with this case, and what were their roles?

2. Where and when did the trial take place? How long did it take? Who was the judge? Was there a jury?

3. What was the background of John Scopes, and why did he allow himself to be used as a test case?

4. Summarize the positions taken by both the defense and prosecution.

5. What was the court's decision? Was it appealed? Was the verdict upheld?

CONCLUSIONS:

1. In your opinion, should evolution be taught as a scientific theory in today's public schools? Defend your answer.

2. There has been a recent movement in parts of the country to gain academic acceptance for the teaching of *creationism* in conjunction with evolution. Using current references (magazine and newspaper articles) to help you, support or reject this idea. Support your opinions with facts.

3. Write a paragraph supporting or opposing the proposition that evolution has indeed taken place. Give evidence to support your point of view.

SUGGESTIONS FOR FURTHER STUDY:

- Research other state laws that have had an impact upon the teaching of evolution. Describe them, including their current status.

- Conduct more extensive research into *creationism*, and the attempts by its supporters to legislate curricula and textbook content. Report on your findings.

- Interview various local religious leaders in your community. What are their views regarding a possible conflict between religious and scientific doctrine? Do they perceive any problems with the teaching of evolution?

DEVELOP YOUR OWN CLASSIFICATION SYSTEM

INTRODUCTION: In order to study the millions of different types of organisms on Earth, biologists have **classified**, or grouped them into five major categories, or **kingdoms**. These kingdoms, in turn, have been divided into smaller, more specific groupings called **phyla** (singular **phylum**). Since the numbers and diversity of organisms is so great, even more successively smaller and more limited groupings are needed to specify an organism. These subdivisions include **class, order, family, genus and species.** This system of classifying organisms proceeds from the general to the specific, much like a person's address. For example, look at the address below, as it would be seen on a letter:

> Jane Doe
> 25 Main Street
> Anytown, New York 11590

How does the post office know where this letter is to be delivered? First of all, the letter belongs in New York state. The state, then, can represent the largest category, or kingdom. But there are still lots of people and places to choose from in this state, so a more specific category, a zip code, is needed. The zip code can represent the phylum. As the chart below illustrates, each piece of information given on the address becomes more and more specific, until only one individual is identified. The specific name of this individual is designated by the genus and species categories. In scientific classification this is also true. All organisms, no matter what they may be called in a local area, have a **scientific name** designated by their genus and species.

EXAMPLES OF CLASSIFICATION SCHEMES

Category	Person's Address	White Oak	Human
KINGDOM	New York	Plant	Animal
PHYLUM	11590	Tracheophyta	Chordata
CLASS	Anytown	Angiosperm	Mammal
ORDER	Main Street	Fagales	Primate
FAMILY	# 25	Fagaceae	Hominidae
GENUS	*Doe*	*Quercus*	*Homo*
SPECIES	*jane*	*alba*	*sapien*

The modern system of classification currently used by biologists is modified from one originally developed in the 1700's by the Swedish naturalist Karl von Linné (Linnaeus). Primarily based on anatomy, this system has been modernized and revised to illustrate evolutionary relationships as much as possible. This particular system of classification, however, is by no means the only one possible. As you know, large groups can be classified in any number of ways. The people in your school, for example, can be grouped according to age, sex, hair color, height, and many other characteristics. The only criteria for a good, workable classification system is that it is consistent and results in the placement of any one individual in one group, only. In this investigation you will gain increased knowledge of classification systems by designing your own. You will also have the opportunity to use an established system to identify various animals.

PURPOSE:

* To gain an understanding of biological classification by designing an original grouping system and dichotomous key.

MATERIALS:

Scissors

PROCEDURE:

Part I: GROUPING GEOMETRIC FIGURES

Illustration 1 contains 20 geometric figures which will represent different "organisms" in a fictitious kingdom called SHAPES. Cut out each "organism," being sure to maintain all its characteristics. Look over the collection and arrange your samples into two "phyla" based on any specific criterion you select. Enter the characteristics of each phylum in the appropriate column, under Trial 1, in Data Table 1. Using only the larger phylum, further divide the group into two distinct "classes." Record these characteristics. In a similar manner, subdivide your largest "class" in two "genera" (plural of genus). If possible, divide the largest "genus" into two "species." Record your groupings on the data table.

Put all the figures back into one kingdom. Can these shapes be grouped differently? To answer this question, repeat the entire classification process using different criteria for your phyla, and other groups. Record the new groupings under Trial 2 on the data table. Answer the following questions on the data sheet:

1. How are the figures in your "phyla" (Trial 1) related? How do they differ?

2. How are the figures in your smallest grouping (Trial 1) related? How do they differ?

3. How are the "organisms" in each of your "classes" related to those in the grouping above them? To the groupings below them?

Part II: USING & CONSTRUCTING A DICHOTOMOUS KEY

If you came upon an organism you had never seen before, could you classify it? How would you go about determining its proper phylum? An extremely useful tool for such identification purposes is called a **dichotomous key**. Such a key is based on a series of two choices, with each decision resulting in another choice, until a result is reached. Such dichotomous keys may give you specific phyla, classes or species, depending on their complexity. The shark key is written to determine the Family to which various sharks belong. Before constructing your own classification key, you will use this one to identify the 10 sharks pictured in Illustration 2. For each animal, begin by making a choice between 1a and 1b. Follow the directions at each choice until you arrive at the Family name for that organism. Write the family name of each shark identified on your data sheet. Above Illustration 2, you are shown some of the anatomical terms used in the key. Study this diagram before you begin.

SHARK KEY	Family or next choice
1.a. Body, as seen from the top, shaped like a kite	Go to 12
b. Body, as seen from the top, not shaped like a kite	Go to 2
2.a. Pelvic fin absent	Pristiophoridae
b. Pelvic fin present	Go to 3
3.a. Six gill slits	Hexanchidae
b. Five gill slits	Go to 4
4.a. Only one dorsal fin	Scyliorhinidae
b. Two dorsal fins	Go to 5
5.a. Mouth all the way at front of snout	Rhincodontidae
b. Mouth on underside of head	Go to 6
6.a. Head expanded with eyes at end of expansion	Sphyrnidea
b. Head not expanded	Go to 7
7.a. Top half of caudal fin same as bottom half	Isuridae
b. Top half of caudal fin different than bottom half	Go to 8
8.a. First dorsal fin extremely long	Pseuodotriakidae
b. First dorsal fin regular length	Go to 9
9.a. Caudal fin extremely long	Alopiidae
b. Caudal fin regular length	Go to 10
10.a. A long point on end of snout	Scapanorhynchidae
b. Snout without long point	Go to 11
11.a. Anal fin absent	Squalidae
b. Anal fin present	Carcharhinidae
12.a. Small dorsal fin present near tip of tail	Rajidae
b. No dorsal fin near tip of tail	Go to 13
13.a. Front of animal with two hornlike appendages	Mobulidae
b. No hornlike appendages	Dasyatidae

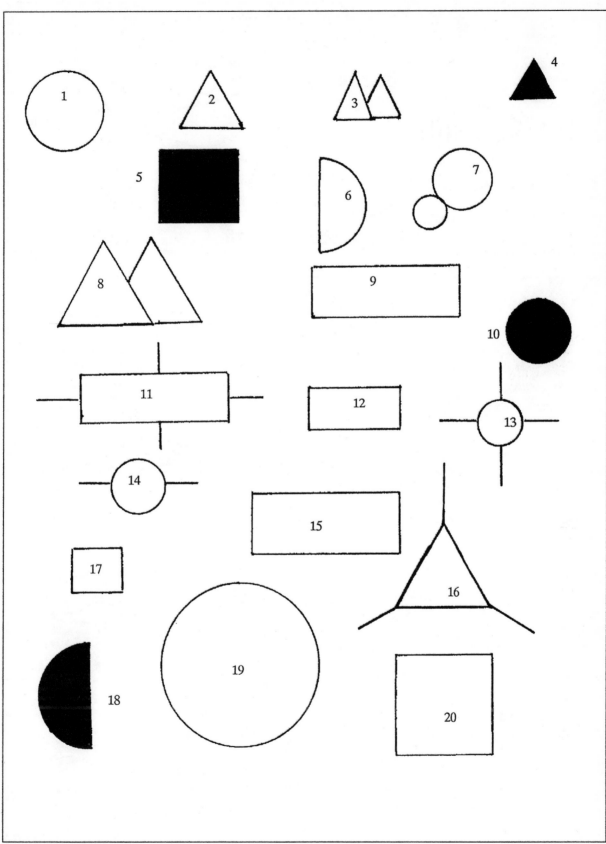

illus. 1: Geometric figures

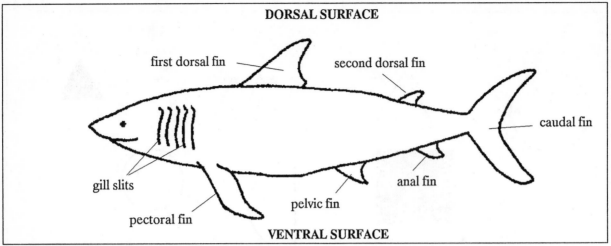

Anatomical terms used in the key

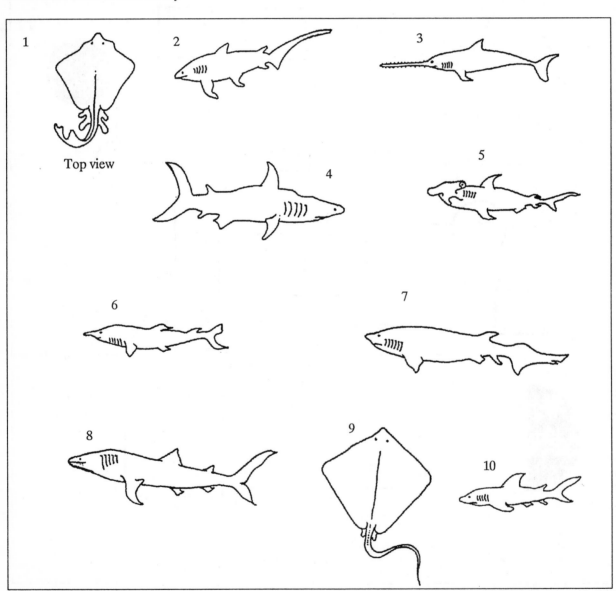

illus. 2: Sharks

You are now going to prepare your own dichotomous key. The organisms you will classify are found in illustration 4. As you can see, these are all fictitious creatures. Assume that they are all in the same kingdom, and use your key to identify their individual phylum group. Use all of the animals shown, with each belonging to a different phylum. On your data sheet, give each of the animals a phylum name, based on whatever unique characteristic that animal possesses. (Animal number 6, for example, might be called "bug-eyed angel fly".) Write your dichotomous key on a separate piece of lined paper and attach it to your lab report. Remember, there is no one *right* key. Several different keys can be made with these animals. The important thing is that your key works and is useful. When you have finished, give your key to another student and see if he/she can use it.

DATA:

DATA TABLE 1: CLASSIFICATION OF SHAPES

Grouping	TRIAL 1		TRIAL 2	
	Characteristics	# of the figures belonging in group	Characteristics	# of the figures belonging in group
Phylum				
Class				
Genus				
Species				

Part I

1. _____

   _____

2. _____

   _____

3. _____

   _____

Part II

Shark Key: Identification of Family names

1._____   6._____

2._____   7._____

3._____   8._____

4._____   9._____

5._____   10._____

Fictitious Animal Key: Identification of Phyla

1._____   6._____

2._____   7._____

3._____   8._____

4._____   9._____

5._____   10._____

CONCLUSIONS:

1. What is meant by the scientific name of an organism? Why are these names useful?

2.   Are two organisms more closely related if they are in the same kingdom or the same phylum? Explain.

3. Using a biology textbook to help you, list the five major kingdoms currently used in scientific classification. Briefly describe the characteristics of each.

4. How many alternatives are there in each step of a dichotomous key?

SUGGESTIONS FOR FURTHER STUDY:

-   Using the information and pictures of leaves below (Illustration 3), construct a dichotomous key to identify each leaf. Using a botanical field guide, look up and record the genus and species names for each of your leaves.

-   Select commonly seen groups of related objects (such as automobiles, hats, or canned foods) and classify them into the major classification categories. Construct a key to their identification. Try your key with some students in your class to see how well it works.

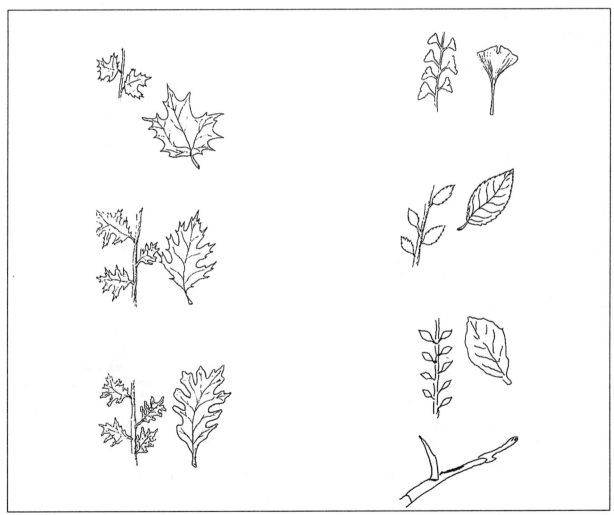

illus. 3: Leaves

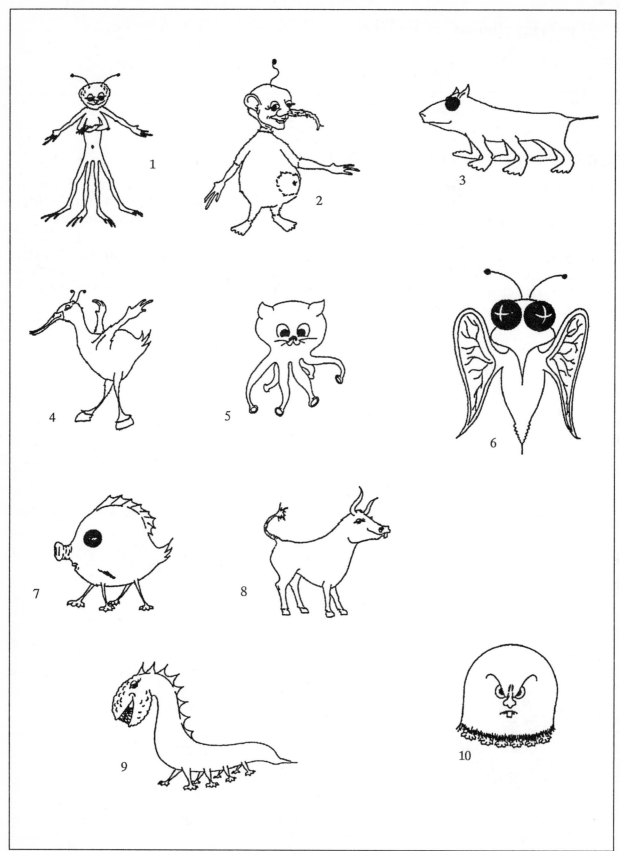

illus. 4: Fictitious animals

ECOLOGICAL SUCCESSION: POPULATION CHANGES OVER TIME

INTRODUCTION: The types of plant and animal populations living in a given ecosystem change gradually over time. Both **biotic** and **abiotic** factors contribute to these changes. As conditions in the ecosystem change, they tend to favor some populations over others. Subsequently, competition causes certain populations to die off while others thrive. Eventually, most ecosystems reach a stable **climax stage**, where populations tend to remain relatively constant as long as no major upheavals in the status quo occur. This gradual change of populations in an ecosystem is known as **ecological succession.**

Most natural succession takes too long for you to observe directly. In this investigation, however, you will set up a small scale pond ecosystem that should exhibit observable succession in a relatively short time period.

PURPOSE:

- How does a pond water ecosystem change over a period of time?

MATERIALS:

Pond water culture	Hot plate
Large beaker	Medicine dropper
Jar with lid	Aged tap water
Dried grass or hay	Thermometer
pH paper	Slides and cover slips
Microscope	Goggles

PROCEDURE:

A. **SET–UP**

To prepare a nutrient medium for your pond organisms, add a handful of hay or dried grass to about 300 ml of aged water in a large beaker. Place the beaker on a hot plate and boil for about 10 minutes, or until the water becomes colored. **Always wear goggles when boiling liquids.** Allow the medium to cool to room temperature (You can speed up this process by immersing the beaker in a large container of cold water).

Pour the liquid, only, into a clean glass jar. Prepare a slide of the water from your jar. Carefully examine the slide under the microscope, looking for signs of life (plant or animal).

1. Did you expect to find any living organisms in the water? Explain.

2. Did your results confirm your expectations? Explain any contradictions.

Innoculate your medium with 3 or 4 droppers of pond culture. Be sure to get representative samples of the pond culture from all levels of the sample; top, middle and bottom. Write your name and the date on the jar.

B. OBSERVATIONS

In order to monitor the abiotic and biotic conditions of your mini–ecosystem, collect the data, indicated below, on the first day, and every other day (or as directed by your teacher), for the duration of the investigation. Record your observations on the Data Table. Between observations, cover the jar **loosely** with the lid and place it in a well–lit area, but not in direct sunlight. Continue making observations for three weeks, or as instructed by your teacher.

DATA COLLECTION:

Abiotic Factors:

1. **Appearance** of the water: observe and describe the color, turbidity and sedimentation, if any.

2. **Odor** of the water: smell and describe.

3. **Temperature** of the water: measure with a Celsius thermometer and record.

4. **pH** of the water: measure with pH paper and record.

Biotic Factors:

To determine the number and kinds of organisms in your pond water, you will prepare one slide as a representative sampling of your culture. First stir your culture gently in order to thoroughly mix its contents.

Place a drop of the water on a slide, cover with a cover slip, and observe under low, then high power of your microscope (remember to adjust the amount of light so that your field is not too bright to see "clear" organisms). Scan your slide in a systematic, organized manner, recording and classifying EACH organism you find. Refer to the reference sheet at the end of this investigation for examples of the types of organisms you might possibly encounter. If you cannot identify a particular organism by name, classify it as a non–green protist, green protist or plant, crustacean, insect, worm or "other." Record the total numbers and names of each type of organism you find on each day of observation.

DATA:

Part A:

1._____

_____

2._____

_____

Part B:

Data Table 1: BIOTIC AND ABIOTIC CONDITIONS OF POND CULTURE

Date	Appearance	Odor	Temp.	pH	Number and Types of Organisms

CONCLUSIONS:

1. What do you think would have happened to your culture if you had put the lid on tightly?

2. What were the first organisms to become abundant in your pond culture?

3. Did you see any evidence of succession in the biotic communities during the course of the investigation? If so, describe the evidence.

4. Were there any changes in the abiotic factors of your pond water during the course of the investigation? Describe and account for any changes you observed.

5. Did your ecosystem reach a climax stage? How can you tell? If yes, what was the climax community? If no, why do you think a climax stage was not reached?

SUGGESTIONS FOR FURTHER STUDY:

- Investigate classic patterns of succession in one or more of the following environments: sandy beach, barren rock, fallow field, volcanic site. Use library materials, museum resources or local land areas in your research. Prepare a report, including pictures or photographs, to illustrate the succession from pioneer to climax community.

SOME COMMON POND WATER ORGANISMS

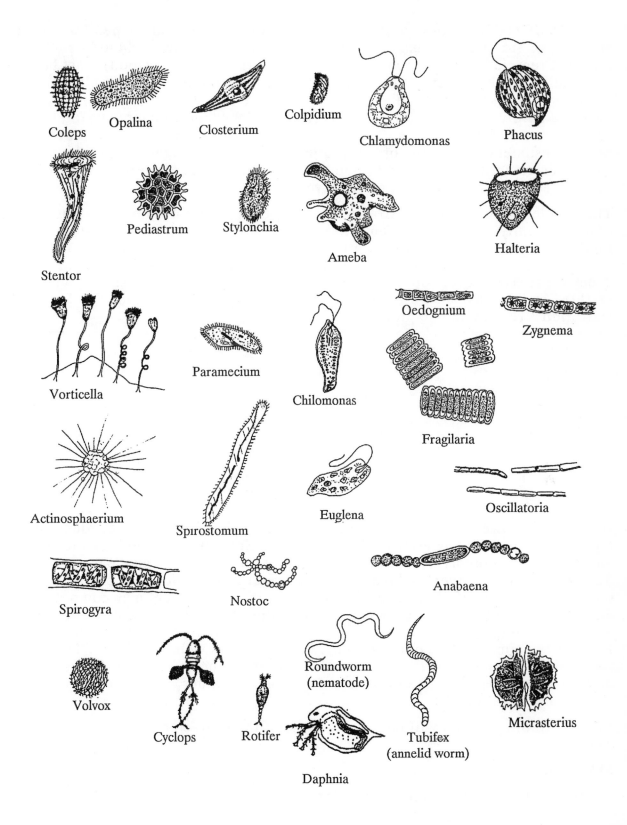

Coleps

Opalina

Closterium

Colpidium

Chlamydomonas

Phacus

Stentor

Pediastrum

Stylonchia

Ameba

Halteria

Vorticella

Paramecium

Chilomonas

Oedognium

Zygnema

Fragilaria

Actinosphaerium

Spirostomum

Euglena

Oscillatoria

Spirogyra

Nostoc

Anabaena

Volvox

Cyclops

Rotifer

Roundworm
(nematode)

Tubifex
(annelid worm)

Micrasterius

Daphnia

WHO EATS WHOM? A STUDY OF A FOOD WEB

INTRODUCTION: In the natural world, all plants and animals depend upon one another for survival. Not only are such substances as oxygen, carbon dioxide, and nitrogen recycled, but also the energy found in organic matter (food) must be distributed to all the **biotic**, or living members of a **community.** The original source of energy for all life on the planet comes from the sun. **Autotrophs,** or **producers,** have the ability to convert the sun's energy into the chemical energy of food. **Consumers,** or **heterotrophs,** can get some of that energy by eating the producers, or other consumers. When any of these organisms die, **scavengers, saprophytes** and **decomposers** eat their remains to obtain their own energy, often returning needed substances to the environment to be recycled.

The simple sequence that illustrates the nutritional relationship between members of a community is known as a **food chain.** (See illustration 1). At the bottom of the food chain, you will always find a producer, like the plant. The consumer that eats the producer, the rabbit in illustration 1, is also known as a **herbivore,** or **first–order consumer.** At the top of the food chain you find another consumer, a **second or third order consumer,** usually a **predator** that hunts for its food. The fox in illustration 1 represents the predator.

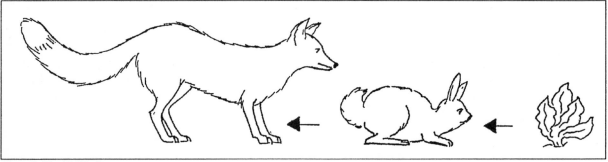

illus. 1: Food Chain

In real life, however, events are not as simple as those shown here. Many other types of plants probably grow in the community. Some may be food for the rabbit, and some may support other consumers. The rabbit, in turn, may eat more than one type of plant. Other herbivores, such as caterpillars, mice, and snails may eat one or more types of plant. Second and third order consumers may compete with each other for the various **prey** animals available. Predators such as owls, hawks, and spiders might choose from a wide variety of first and second order consumers. Still other animals, the **omnivores,** might eat plants as well as animals. This complex pattern of nutritional dependence is closer to the reality that exists in any natural **ecosystem,** and is known as a **food web.** A rather simplistic food web is shown in illustration 2.

In this exploration, you will investigate some of the characteristics of various food webs.

MATERIALS:

Biology text or other reference material.

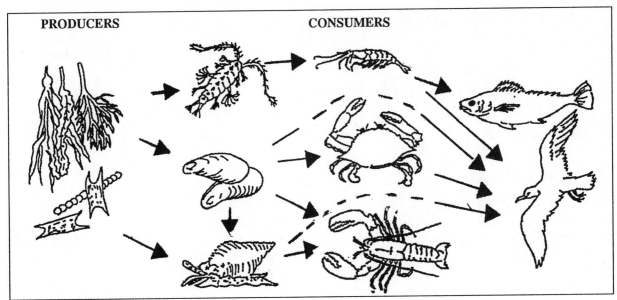

illus. 2: Food Webb

PROCEDURE:

Part I: IMBALANCE

In one natural Arizona community, deer made up a large percentage of the first order consumers, grazing on wild grasses. Their predators, wolves, were the only second–order consumers in the community. As more and more people settled in the area, more and more wolves were killed. (The wolves were mistakenly perceived as a dangerous threat to the uneducated settlers, who shot them on sight). The graph below shows the population curve for deer in this community. Based on this information, answer the following questions on your data sheet:

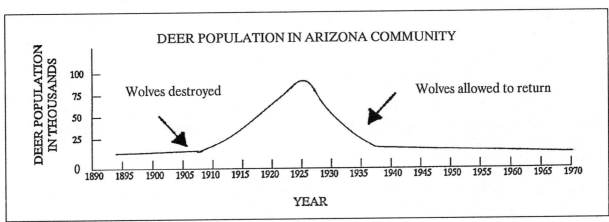

1. Why did the deer population increase between 1908 and 1923?

2. How can you account for the sudden decline in deer population around 1925?

3. If rabbits were present in this community, how do you think their population would be affected by this data? What would their population curve look like?

4. Why do you think the Arizona settlers allowed the wolves to repopulate the area in 1935?

5. How do you explain the deer population curve from 1935 to 1970?

6. What is the normal "carrying capacity" of this land in terms of the deer population that the area can support?

7. If you were to draw a curve showing the population of native grasses in the community between 1900 and 1970, what would it look like?

8. How might the population of a scavenger, such as a vulture, fluctuate during the period from 1915–1940? Explain.

Part II: PYRAMIDS

When producers convert the sun's energy into food energy, they use some of it for daily functions, store some, and use some to build new plant tissue. When a herbivore, such as a cow, eats the plant, does the cow get 100% of the plant's energy? Will one plant, such as alfalfa, support one cow? Of course not! Much of the plant is of no use to the cow. A single cow may need millions of plants in order to survive. The small amount of usable energy in each plant is then used to support the cow's daily activities. A very small fraction of the energy is stored, or used to make "more cow." As you can see, the amount of energy available at any level in the food web **decreases** as you go higher and higher into the chain. The same is true for numbers and mass of organisms. This concept is often referred to as the **pyramid of energy, mass or numbers.** The chart below relates to a simple food chain including alfalfa, cows and humans. Use it to answer the questions that follow. Put your answers in the appropriate spaces on the data sheet.

Organism	Numbers	Mass	Available Energy
young human	1	50 kg	.06 %
beef cow	4	1,000 kg	7.00%
alfalfa plant	20 million	8,000 kg	100%

1. Fill in the pyramids on the data sheet so that the three organisms are placed in their correct levels.

2. How many alfalfa plants are needed to support the amount of beef eaten by one young human in a year?

3. If 2 alfalfa plants can grow in 6.25 cm^2, how much alfalfa farmland is needed to support one cow? Four cows?

4. The mass listed in the table refers to the total mass in the food chain for each organism during a year. What *percent* of the *total* mass is accounted for by the producers? The human?

5. If there was a drought and 50% of the alfalfa crop was ruined, how would that affect humans?

Part III: FOOD WEB

The types of organisms found in a given food web depend upon the type of **ecosystem** you are investigating. Various ecosystems include:

> fresh water pond, temperate deciduous forest, marine (oceans), prairie or grassland, desert, beach, coniferous forest, swamp.

Select one of these ecosystems, or another with teacher approval, and research the types of plants and animals that live there. You may choose to investigate the ecosystem in which you live, and determine the types of plant and animal life by direct observation. On the data sheet, list at least 10 organisms found in your chosen ecosystem. In the space provided, draw a representative food web that includes all the organisms listed. Base your answers to the following questions on your food web.

1. Which organisms are producers in your ecosystem?

2. Which organisms are first–order consumers?

3. Which organisms are second–order consumers?

4. Which organisms are predators?

5. Which organisms would be found in the most numbers?

6. Which organisms would contribute the *least* mass to the food web?

7. Describe the non–living, or **abiotic** factors that contribute to the general make–up of your ecosystem.

8. Using 4 of the organisms in your web, construct a pyramid of energy, placing each of the 4 organisms in their appropriate level.

DATA:

Part I

1._____

2._____

3._____

_____

4._____

5._____

_____

6._____

7.

Wild
grasses ↑
population

Years →

8._____

_____

Part II:

1.

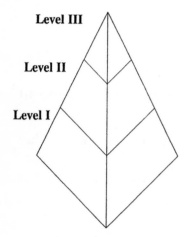

Pyramid of energy

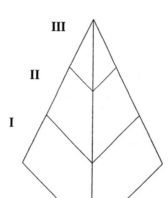

Pyramid of mass

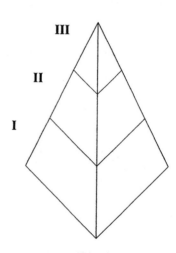

Pyramid of numbers

2._____

3._____

4._____

5._____

_____

Part III

ORGANISMS IN YOUR ECOSYSTEM:

1. _____

2. _____

3. _____

4. _____

5. _____

6. _____

7. _____

8. _____

9. _____

10. _____

DIAGRAM OF YOUR FOOD WEB:

1. _____

2. _____

3. _____

4. _____

5. _____

6._____

7._____

8.

CONCLUSIONS:

1. Certain human activities, such as lumbering, road building, and farming, cause the destruction of plant or animal species in a given community. Other than concern for that particular species, why should we make every effort to disturb the natural populations of a community as little as possible?

2. You saw in part II of this activity that it requires quite a large area of land to support the producers in a given food chain. In light of this information, how might you account for the decline in population of certain predator species, such as tigers, bobcats and ospreys, in areas with increasing human development?

3. Why is there a loss of energy as you move up in a food chain?

4. Distinguish between an ecosystem and a community.

SUGGESTIONS FOR FURTHER STUDY:

- Design and establish your own self–sustaining aquatic ecosystem. This can be accomplished by planning, in advance, a desirable aquatic food web, and setting this up in a closed aquarium tank. You will need aquatic plants as producers, organisms, such as snails and small fish, as first order consumers, and larger fish as second–order consumers. Bactria, to act as decomposers, can be obtained from bottom mud or sand in a local freshwater pond. Local pond water can also supply microscopic plankton, both autotrophic and heterotrophic. Examine samples of your water to see if these organisms are present.

  Research the various nutritional needs of the available fresh water organisms before you begin your planning. Be sure to take into account the numbers and masses of the organisms that are required at each trophic level. Keep your food web relatively simple while ensuring that the organisms at the top of the chain will get enough energy to survive. Plan on locating your ecosystem near a window that receives sun during a large portion of the day. If this is not possible, artificial light may be substituted.

  When you have planned your ecosystem and decided upon the organisms to include in your food web, obtain teacher approval. Set up your tank and make careful observations on a regular basis. Collect data on the numbers of organisms over a period of several months. Report your results.

A STUDY IN HUMAN POPULATION GROWTH

INTRODUCTION: The world's population recently reached 5 billion. To get an idea of how big a number that actually is, think of it as money. If you had five billion dollars, how long would it take to spend it at the rate of $1,000 a day? It would take 5,000,000 days, or 13,700 years! More alarming is the fact that the population is increasing at an ever increasing rate. At its present rate of growth (1.7% year), the population of the world is expected to double in 40 years! Where will 10 billion people live? How will we produce twice as much food as we do now? Where will twice as much electricity come from? Many people are now realizing that the capacity of the planet to support human life is not unlimited. The population pressures are made even worse by the fact that the people and resources of the planet are distributed unevenly. Asia, for example, has approximately 58% of the population, but only 20% of the land mass. North American has approximately 6% of the world's population, but uses over 40% of the energy. These, and other inequities, add further stress to an increasingly crowded planet.

In the 1700's, the British economist Thomas Malthus was one of the first people to warn about the dangers of overpopulation. He illustrated that populations increase in a **geometric progression**, where five becomes 10, 10 becomes 20 and 20 becomes 40. This type of growth, he and others contended, would soon overcome the food supply, resulting in famines, wars and other disasters. In the 1960's, Dr. Paul Ehrlich again took up the campaign to curb world population growth. Although everyone is not in agreement with the predictions of some population control advocates, the need to slow the rate of population growth is becoming more and more accepted. In China, for instance, the government rewards (or even requires) one–child families. **Zero population growth** (ZPG) is a movement which strives to adjust the worldwide birthrate to equal the death rate, through education, so that no population growth occurs. Most countries, however, are quite far from this goal at the present time.

PURPOSE:

- What can be learned and predicted from the pattern of human population growth?

MATERIALS:

Graph paper

PROCEDURE:

Part I: Exponential Growth

A. In order to understand the nature of exponential increases, try the following riddle:

> A father complained that giving his son an allowance of $10 per week was too excessive. The clever son then suggested an alternative. He would agree to get no more allowance if his father would give him 1 penny on the first day of the month, and double the amount every day

for the rest of the month. The father, therefore, would give the son 1 cent on the first day, 2 cents on the second, 4 cents on the third, eight on the fourth, and so on for 30 days. If the father agrees, did the son make a good deal? On your data sheet record the amount of money the son would have at the end of his 30–day proposal. How much would he have accumulated if he took his regular allowance?

B. As you can see, exponential growth can increase quite rapidly. This is the type of growth normally exhibited by natural populations. In nature, however, the growth is never unlimited. At some point a natural population will run out of some of the materials necessary for its survival. These limiting factors are usually food, living space or raw materials needed to support life. An excellent way to visualize this growth pattern is to construct a graph of population growth. The data below were collected during an actual investigation of the growth of bacteria in a test tube. (Bacteria are often used in population studies because they reproduce quite rapidly, by asexual means).

A small amount of bacteria were transferred into a tube with nutrient media. The culture was incubated at a temperature suitable for the growth of this particular bacteria. Samples were taken at one–hour intervals and population counts were made. The data is supplied on the table below:

GROWTH OF A BACTERIAL POPULATION

Time (hr.)	Number of Organisms	Time (hr.)	Number of Organisms
0	1,500	8	76,000
1	2,000	9	77,000
2	4,000	10	74,000
3	7,000	11	67,000
4	14,000	12	58,000
5	30,000	13	45,000
6	59,000	14	35,000
7	71,000		

On a separate sheet of graph paper, graph the data shown in the table. Put time (in hours) on the horizontal axis and population on the vertical axis. Examine the curve you drew and answer the following questions on your data sheet:

1. During which hours was the bacteria population growing at a faster rate than it was dying off?

2. During which hours were they maintaining a relatively stable population, where the birth rate approximately equaled the death rate?

3. During which time was the population in a death phase, where more individuals were dying than were being "born?"

C. The human population can also be studied using population data and growth curves. Since humans reproduce quite differently than bacteria, our data will not accumulate as quickly as that in section B. You will see, however, similarities in the growth patterns of both populations.

The table below shows selected data for world population since the 1600's. On a separate sheet of graph paper, draw a graph of this data. Plot years (in 25 year intervals) on the horizontal axis, and world population (in billions of people) on the vertical axis. Examine your graph and answer the following questions:

1. In what ways is this graph similar to that of the bacterial population?

2. In what ways is it different?

3. How long did it take for the world's population to double from .5 billion to 1 billion? From 1 to 2 billion? From 2 to 4 billion?

4. What is the current predicted doubling time for our population?

GROWTH IN WORLD POPULATION (1600-2026 (predicted))

Year	Estimated Population (in billions)	Year	Estimated Population (in billions)	
1600	.5	1978	4.2	
1700	.6	1981	4.5	
1800	.8	1986	5.0	
1830	1.0	1992	*5.5	
1900	1.6	2000	*6.2	
1930	2.0	2020	*8.6	
1960	3.0	2026	*10.0	*Projections
1975	4.0			

Part II: WORLD POPULATION DATA

A. The human population problem is not one of mere numbers. The way in which the population is distributed around the world is also of great concern. Some regions are growing much faster than others. Some regions have more land or resources than others. These, and other unbalances lead to the establishment of "have" and "have not" societies, often causing social unrest and political conflicts. The table on the next page shows some population, and related data, for six regions of the world. Study the table and use the data to answer the questions below:

1. Identify the regions with the largest and smallest 1984 populations. Now look at the amount of arable land per person in each of the regions. How do these two sets of data compare? What does this tell you about the ability to feed the populations of each of these regions?

2. What are the three fastest growing regions in the world? In how many years will each double their populations?

POPULATION DATA FOR VARIOUS REGIONS OF THE WORLD

Region	1984 Pop. (millions)	Projected 2000 Population (millions)	Growth Rate (%)	Doubling Time (years)	Arable Acres Per Person
NORTH AMERICA	262	297	0.7	99	2.20
SOVIET UNION	274	316	2.1	68	2.10
LATIN AMERICA	357	562	2.6	30	1.10
EUROPE	491	510	0.3	208	0.71
AFRICA	531	855	2.9	24	0.85
ASIA	2,782	3,680	1.8	38	0.40

3. Which region has the most stable population? How long will it take this region to double its population?

4. Two important types of data are missing from the table above. These are energy consumption and per capita gross national product. The former refers to the total amount of the world's energy resources used by a particular region, while the latter refers to the amount of money, per person, produced in each region. Which world region do you think leads in each of these categories? Which regions are probably at the lowest end of the scale in each category? How do your "highest" and "lowest" regions compare in population and population growth?

B. One of the best ways to slow population growth is to limit the size of families. Some countries are currently making major efforts to promote birth control and limit the number of children born. These efforts, however, are meeting with limited success in most areas. The illustration on the next page shows the relationship between population growth and the average number of children per family. Examine the chart and answer the following questions on your data sheet.

1. After four generations, how many **times** greater is the population from a three–child family than from a one–child family?

2. How much more food and water, in percentages, will a three–child family require after four generations as compared with a one–child family tradition?

3. Calculate the number of people that will be produced after four generations of a four–child family tradition.

Relationship Between Population Growth and Average Family Size Patterns		
ONE–CHILD FAMILY	TWO–CHILD FAMILY	THREE–CHILD FAMILY
1 person = 1 generation	1 person = 1 generation	1 person = 1 generation
1 person = 2 generations	2 people = 2 generations	3 people = 2 generations
1 person = 3 generations	4 people = 3 generations	9 people = 3 generations
1 person = 4 generations	8 people = 4 generations	27 people = 4 generations
4 people total	*15 people total*	*40 people total*

To visualize the significance of two vs. three child families, graph the data below, which project the population of the United States. Put "years" on the horizontal axis, and "population" (in millions) on the vertical axis. Plot the curves for one child families on the same axis as the curve for two–child families. Label each curve.

On your data sheet, write a summary of the meaning or significance of your graph.

UNITED STATES POPULATION AND PROJECTIONS

Year	Population in Millions	
1870	50	
1890	75	
1920	100	
1950	175	
1968	200	
	Estimated Projections	
	Two Child Families	**Three Child Families**
1995	280	300
2013	295	400
2030	310	500
2050	330	700
2070	360	1000

DATA:

Part I:

A. Regular allowance total after 30 days_____

    "Exponential" allowance after 30 days_____

B. 1. Hours representing growth phase_____

    2. Hours representing stationary phase_____

    3. Hours representing death phase_____

C. 1._____

    2._____

    3._____

    _____

    4._____

Part II

A. 1. _____

    _____

    2._____

    _____

    3._____

    4._____

    _____

    _____

B . 1._____

    2._____

    3._____

    4. Summary of graph:

CONCLUSIONS:

1. What are some of the factors that could have contributed to the death phase in the bacterial populations described in Part I?

2. Compare your graph of the human population in Part I to that of the bacteria. What phase of the bacterial growth curve does the current human population curve most resemble? What do you predict will happen to the human population if it keeps increasing as it is?

3. The bacteria were grown under strict conditions of limited food and space. If the population counts were continued, the numbers would probably go down to zero. Many people do not believe the same would be true of the human population. Why not?

4. From the information given in Part II: A, *World Population Data Table,* summarize what is meant by "have" and "have not" nations.

5. Why do you think that population control measures, such as limiting family size, are not meeting with much success in poorer, less developed regions of the world?

6. What factors have contributed to the world–wide "population explosion" of the past 200 years?

7. What are some of the problems that the world will be facing as the human population continues to escalate?

SUGGESTIONS FOR FURTHER STUDY:

- Read Dr. Paul Ehrlich's book *The Population Bomb* (Ballantine Books, New York. 1968). Summarize his main points in a written report. Have any of his concerns or predictions changed since he wrote the book in 1968? Have any new problems arisen?

- Conduct a survey in your school to determine the students' attitudes toward population controls and policies. One method to accomplish this is to prepare a response sheet with various statements to which the students agree or disagree, on a scale from one extreme to the other. Prepare your own statements, or use some of the examples below. Get approval from your teacher before proceeding with your project. Report your results in a scientific fashion. Sample opinion statements include:

  a. Our country should adopt a strong population policy, including yearly limits on immigration and rigid goals on population size.

  b. In the event of too much crowding, all public lands, including national parks and wildlife preserves, should be turned over to developers for housing.

  c. Families with more than two children should be penalized by paying higher taxes.

  d. At some time in the future we are likely to deplete essential resources.

THE PROBLEM OF ACID RAIN

INTRODUCTION: Before the Industrial Revolution, the average pH of rain falling on the earth was slightly acidic, approximately 5.5 – 6. Today, by contrast, the rain falling over eastern North America and western Europe has an average pH of 4.0 – 4.5. This represents more than a 10–fold increase in acidity. Worse still, in some parts of Scotland precipitation with a pH of 2.4 has been recorded! What is causing this **acid precipitation**? What are its effects? How can it be corrected?

These and other related questions are currently being debated by scientists throughout the world. But the potential problems of acid deposition are not only ecological concerns, but are political and economic as well. Who is responsible for the acid precipitation falling in Scandinavia or Canada or New York? Who should pay for any preventative or corrective measures deemed necessary? These problems, and others, are not going to be solved overnight. Conflicts between technological "progress" and environmental stability have been growing as rapidly as our population, and compromise has been difficult. The acid precipitation dilemma merely is adding to this debate.

In this activity, you will research the basis of acid precipitation and its consequences. From your findings, you will be asked to formulate possible remedies, just as the scientific and political leaders of the world are currently attempting to do.

PURPOSE:

- What are the causes and consequences of acid precipitation?

MATERIALS:

pH paper or meter
Two petri dishes
Filter paper
Granite chips
Reference books/articles

Rain or snow samples
Marigold seeds
Marble chips
Acid solution – pH 3
Small plastic container

PROCEDURE:

Part I: WHAT CAUSES ACID PRECIPITATION?

A. Simply put, acid precipitation (rain, snow, smog, fog, etc.) is caused by **sulfur dioxide** (SO_2) and **nitrogen oxides** (NO_x) in the atmosphere. Photochemical reactions then produce sulfuric and nitric acids. As precipitation falls, the acids are carried down to earth as acid precipitation. This explanation, however, leaves many questions unanswered. Using reference materials from your classroom, school and public libraries, answer the following six questions on a separate sheet of paper:

1. What are some *natural* sources that contribute sulfur dioxide and nitrogen oxides to the atmosphere?

2. What are the major man–made *(technological)* sources that are adding these gases to the atmosphere?

3. In what quantities are the gases being added to the atmosphere by the natural and man–made sources?

4. What region of North America contributes most of the sulfur dioxide emissions to the atmosphere?

5. What region of North America receives the most acid precipitation?

6. What are the prevailing wind currents in North America?

B. To determine if there is acid precipitation in your area, you simply have to collect rain or snow samples and measure their pH. Place a plastic container outside your school on a rainy or snowy day. Collect a small sample of the precipitation and bring it back to the classroom. Using a pH meter or pH paper, determine the pH of the precipitation. Record your results on the data page. How acidic is the precipitation in your area?

Part II: WHAT ARE THE EFFECTS OF ACID DEPOSITION?

Acid precipitation affects lakes, streams and forests. Many factors, however, combine to determine the severity of these effects. Natural **buffering** systems in the bedrock and soil may help some areas recover from increased acidity. But, unfortunately, the mountainous areas receiving most of the acid precipitation are often those with the least ability to recover.

Perform the following activity, which will help you determine the buffering ability of two types of rock; granite (a silicate) and marble (a carbonate).

a. Fill the bottom of one test tube with about 1 cm of marble chips. Fill another with an equal amount of granite chips. Label each tube appropriately.

b. Pour about 10 ml of acid solution into each test tube. Shake. Record the pH of the solution at the start of the experiment in Data Table 1.

c. At 15 minute intervals, record the pH of the solution in each tube for the next hour. (Continue with the investigation while you are waiting between readings).

d. On your data sheet, summarize the buffering power of marble and granite.

Using available references, answer the following questions, on a separate sheet of paper, about the effects of acid precipitation.

1. Why are mountain lakes and streams, such as those in the northeastern United States, more susceptible to damage by acid precipitation?

2. In what ways does a decrease in pH harm an aquatic ecosystem such as the type described in question 1?

3. What are the current figures regarding the destruction of lakes in the Adirondack and Appalachian Mountains of northeastern America?

Forest ecosystems are also being affected by acid precipitation. Although most forests have the ability to purify much of the precipitation that falls on them, there does seem to be a limit to their buffering capacity.

4. On your separate paper, describe several ways in which acid precipitation may harm forest ecosystems, including its affect on the soil, bacteria, foliage and tree growth.

To determine the possible effects of acid precipitation on the germination of seeds, perform the following activity:

e. Line the bottom of two petri dishes with several layers of filter paper. Soak the papers in one of the dishes with distilled water. Label this dish C, for control. Soak the papers in the other dish with the pH 3 acid solution. Label this dish A, for acid.

f. Sprinkle 20 marigold seeds into each of the dishes. Place the covers on the dishes on an angle, so that air can still circulate around the seeds, but moisture will not readily evaporate. Place both dishes in a warm, undisturbed area.

g. Observe the dishes each day for the next week, looking for evidence of seed germination. Record the number of seeds that germinate in each dish, each day, in Data Table 2. Be sure to keep the paper in each dish moist by adding either water or acid, as needed. Remove any seeds that have germinated as soon as you count them so that you don't count a seedling more than once.

h. On your data sheet, summarize the effect of acid on seed germination.

There is also growing evidence that acid precipitation is damaging crop plants, such as bush beans, soybeans, apples, and others. Using your reference material, answer the following:

5. What evidence is there that crops are being damaged by acid precipitation?

Another consequence of acid precipitation is the destruction of buildings and statues in many cities of the world. In the U.S. alone, EPA specialists estimate the cost to repair or replace structures damaged by acid rain at over **$5 billion per year!**

6. What evidence is there to suggest that buildings and statues are being damaged/destroyed by acid precipitation?

7. Is there any evidence or concern that acid precipitation will affect human health?

Part III: WHAT CAN BE DONE ABOUT ACID DEPOSITION?

Using your reference material, answer the following questions:

1. How might adding lime to lakes help the acid problem? Why do many environmentalists oppose this measure of control?

2. How might sulfur dioxide emissions be reduced?

3. Any measures mentioned in answer to question 2 will cost money. Who should pay for these measures?

4. What do industrial groups, such as the Electric Power Research Institute and the American Petroleum Institute, argue about possible solutions to the acid rain dilemma?

5. Is there any evidence of cooperation between governments, such as the U.S. and Canada, in addressing the problems of acid precipitation?

6. What types of research are currently being conducted in the area of acid precipitation?

DATA:

B:

Part I: pH of local precipitation: _____

Is your area affected by acid rain or snow?_____

DATA TABLE 1: BUFFERING ABILITY OF MARBLE AND GRANITE

pH	0 minutes	15 minutes	30 minutes	45 minutes	60 minutes
MARBLE					
GRANITE					

Part II:

d. Summary of the buffering ability of granite and marble:

_____

_____

DATA TABLE 2: SEED GERMINATION IN ACID AND DISTILLED WATER

Dish	Day 1	2	3	4	5	6	7	% Germination
"C"								
"A"								

h. Summary of the effects of acid on marigold seed germination:

_____

_____

CONCLUSIONS:

1. Why is acid precipitation much more of a problem today that it was 100 years ago?

2. Why are areas such as the northeastern United States, Scandinavia and south-central and southeastern Canada affected so much by acid precipitation?

3. Suppose that a panel of scientists recommended that in order to reduce nitrous oxide emissions, all private use of automobiles should be restricted. How do you think the public would respond to this recommendation? How do you feel about it personally? How much do you think people would, or should, "sacrifice" in order to alleviate the problems caused by pollution?

4. Do you believe there is adequate scientific evidence showing the relationship between sulfur dioxide emissions and damage caused by acid precipitation? Explain.

5. In light of your answer to question 4, what action, if any, would **you** propose at this time. Explain.

SUGGESTIONS FOR FURTHER STUDY:

*   Contact your local political leaders to determine their stand on the issues surrounding acid precipitation. Are they aware of the issues? What are their thoughts on the issues? Report back to your class, and evaluate the responses you were given.

*   Investigate the work of F.H. Bormann at the Hubbard Brook Experimental Forest in New Hampshire. How has his work contributed to our knowledge about the effects of acid precipitation? How does his work relate to the crisis currently being investigated in the Black Forest of Germany?

ACID RAIN: THE BITTER DILEMMA

JENNIFER ANGYAL

From the Audiovisuals/Scientific Publications Department,
Carolina Biological Supply Company, Burlington, North Carolina 27215

The autumn rains falling over eastern North America bathe our continent in acid. The raindrops splattering pleasantly against your windowpane are likely to be a dilute solution of nitric and sulfuric acids; sometimes, they may be as acidic as vinegar. Even if you live far from centers of industry and human congestion, your rainfall may be a bitter potion. The problem is spreading from the industrialized east into the open spaces of the west: high in the Colorado Rockies, not far from the Continental Divide, acid rain falls over the remote Como Creek watershed.

The acid rain that sterilizes our lakes and injures our forests is formed when oxides of sulfur and nitrogen are washed from the air by rainwater, forming a solution of sulfuric and nitric acids. Excess sulfur and nitrogen compounds pour into the atmosphere from the burning of fossil fuels. But if we have thus created acid rain, why can't we control it?

The story of acid rain serves to illustrate the fundamental conflicts that keep mankind trapped in the downward spiral of environmental deterioration. These conflicts are of two kinds. The first clash occurs between the need for immediate action and the need for complete scientific information to guide that action. Paradoxically, action without knowledge may be catastrophic, but waiting to act until we know everything may be catastrophic, too. For example, two or three decades ago industry began building taller smokestacks to alleviate local air pollution. This solution worked, but, as we shall see, it contributed to the regional problem of acid rain. Our knowledge was too incomplete to anticipate this disastrous result.

Today, on the other hand, acid rain is almost certainly damaging our lakes and forests, but, because its causes and effects are incompletely understood, the best course of action is unclear. By the time we understand the problem fully, however, the damage may be irreversible.

The second dilemma involves a conflict between apparently incompatible values: pollution is usually the unfortunate side effect of some activity which truly benefits society. Coal, a high-sulfur fuel, contributes more heavily to the acid rain problem than does oil; but burning coal affords us the political independence of using our own domestic fossil fuel reserves. Controlling sulfur emissions, whether by burning expensive foreign oil or by applying technology to the combustion of coal, is expensive. Inevitably, the costs are borne by the taxpayer and the consumer. But in the long run, may not pollution itself be more expensive?

These two conflicts—the need for action versus the need for knowledge, and the cost of pollution versus the cost of controlling it—are closely interrelated. Learning what we must know to make informed decisions is itself an expensive proposition: in August of last year, President Carter called for the allotment to acid rain research of $10 million annually.

A closer examination of the acid rain problem, its history and its possible outcomes, is instructive not only in itself but as an example of the kinds of dilemmas that we face in tackling any of our environmental problems.

COAL AND THE LOCAL POLLUTION PROBLEM

During the Industrial Revolution, the proliferation of coal-burning industries spewing soot into the air soon created serious local pollution problems. Soot consists of airborne particulate matter, or fly ash. Clinging to the fly ash particles is sulfuric acid, formed from water and sulfur dioxide. The colorless sulfur dioxide gas is generated during the combustion of high sulfur fuels such as soft coal. Nitrogen oxides, produced in the same way, color the air with a brown haze. Together, these emissions create the atmospheric condition known as "London smog."

The gaseous oxides of nitrogen and sulfur irritate the eyes, nose, and throat. Like carbon monoxide, nitrogen oxide combines with hemoglobin far more readily than oxygen does, and at sufficiently high concentrations it can cause death by asphyxiation. It also lowers resistance to infections and increases susceptibility to various respiratory diseases. Inhalation of sulfuric acid, clinging to particulate matter or suspended as a mist in the air, irritates the respiratory tract and damages the cilia in the lungs. Pulmonary hemorrhage and permanent lung damage have been induced in laboratory animals by exposure to sulfuric acid mist. Methylated sulfates, which are organic derivatives of sulfur oxides, are both mutagenic and carcinogenic. One recent study shows that these compounds are present in significant amounts in the respirable particulate matter generated by coal-fired power plants.

However, it did not require very sophisticated understanding of its hazards to know that the irritating haze of coal smoke was unhealthy. And it soon became apparent that it was unhealthy not only for human beings but for other living things as well. By the turn of the century, emissions from smelters processing ore at Copper Hill, Tennessee, had completely destroyed 28 km² of deciduous forest. Examples of damage to local ecosystems by ore smelters, which are among the major users of coal, are too many to enumerate, but one of the most dramatic is in Sudbury, Ontario: a single smelter complex spews out 1 to 3 percent of the total worldwide sulfur emissions every year. Nineteen hundred km² of forest suffer damage under this assault.

THE LOCAL PROBLEM GOES REGIONAL

What could be done about the local pockets of industrial air pollution that contaminated cities and destroyed forests? One simple and obvious solution was to send the poisonous emissions high into the sky to be swept away by the winds. Following several especially severe incidents of London smog in the 1950s, Great Britain adopted a policy of building taller and taller smokestacks to alleviate the problem, and other nations soon followed suit.

The tall smokestacks brought a welcome breath of fresh air to industrial areas. The oceans of air moving hundreds of meters above the earth's surface were a convenient dumping ground that carried the noxious wastes well out of sight—and out of mind. Who could have predicted that fish in Sweden would begin to die because the smokestacks of western Europe towered higher and higher?

But the oxides of sulfur and nitrogen deposited high in the atmosphere had to go somewhere; they could not disappear. Carried aloft on prevailing winds for several days, the oxides have time to undergo photochemical reactions which convert them into acids. Rain and snow then wash the acids from the air. Sweden and Norway have the misfortune to be in the path of the southerly winds blowing from western Europe—and carrying with them Europe's acid wastes. In 1959, a Norwegian Fisheries Inspector was the first to attribute declining fish populations to the acidification of Scandinavian lakes and streams. The region of acid precipitation has steadily spread, and today acid rain falls over most of northwestern Europe.

Acid rain was also falling over much of the northeastern United States by the mid '50s. Since that time, rainfall in the northeast has become still more acidic, and the region receiving acid precipitation has spread to include much of North America east of the Mississippi.

One recent report shows increasingly acid precipitation falling over a remote area in the Colorado Rockies. The nearest center of human population is the Fort Collins-Denver-Boulder area lying 20 to 50 km to the east. But the prevailing winds over Como Creek blow from the northwest, so they may be bearing their bitter cargo from as far away as Salt Lake City, 600 km to the west, or from some still more distant source. Acid rain respects no human boundaries.

SOME MAGNITUDES

Just how acid is acid rain? How much blame for it can be placed on human activity? Many of the figures that follow are rough estimates at best, but there is enough agreement among the estimates from various sources to suggest the magnitude of the problem.

One common measure of acidity is the pH scale; which ranges from 0 to 14. Pure water is neutral at pH 7, and lower values represent acids. Each *unit* change in pH represents a *tenfold* change in acidity; thus a solution with pH 4 is 10 times as acidic as one with pH 5.

Normal rainfall is not pure water, and its pH is not 7. Water vapor in the atmosphere reaches an equilibrium with gases in the air, among them carbon dioxide. Dissolved in water, carbon dioxide produces a solution of carbonic acid. This weak acid dissociates slightly, lowering the pH to 5.6—the lowest pH expected for rainfall in an unpolluted atmosphere. However, the actual pH of normal rain and snow is often above 6 because of the presence of airborne dust from slightly alkaline soils.

Not surprisingly, there are few records of rainfall chemistry from the days before anyone suspected a problem. However, rain and snow that fell before the Industrial Revolution have been preserved in glaciers and ice sheets; such "fossil rain" generally has a pH above 5. The rain falling today over eastern North America and western Europe has an average annual pH of 4 to 4.5. Individual storms may be much more acidic—in 1974 a storm over Scotland dumped rain as acid as vinegar, with a pH of 2.4.

Where does all the acidity come from? Scientists estimate that about 60 percent of the acidity of rainfall is due to sulfur dioxide in the atmosphere, while the remaining 40 percent is due to nitrogen oxides. It is exceedingly difficult to estimate just what proportion of these atmospheric oxides is generated by human activity, but it is probably of the same order of magnitude as that produced by natural sources. In the natural biogeochemical cycles of sulfur and nitrogen, these elements spend only a short time as atmospheric oxides. The largest reservoir of nitrogen is the gaseous N_2 in the air, while most of the earth's sulfur is tied up in soils and sediments.

Natural sources of atmospheric sulfur include weathering of the earth's crust, airborne sea salts, volcanoes, and the activity of anaerobic sulfur bacteria in lakes, bogs, and seashores. One report estimated in 1971 that the rate of sulfur introduction from fossil fuels was roughly equal to that from anaerobic areas and volcanoes combined. The huge smelter complex in Sudbury, Ontario, is thought to have emitted as much sulfur every year for the last decade as all of the world's volcanoes. The U. S. Environmental Protection Agency estimates that 30 to 50 percent of the sulfur dioxide in this country's air comes from human activity. In southern Sweden, as much as 70 percent of the atmospheric sulfur may be from human sources—most of them outside of Sweden—and in the coal areas of western Europe, the estimate rises to 90 percent.

The actual quantity of sulfur dioxide that the U. S. dumps into the atmosphere annually is on the order of 20 million tons, although estimates vary somewhat. Of this, roughly two-thirds is produced by the combustion of fossil fuels in power plants. The EPA estimates that U. S. power plants alone produced 18.6 million tons of sulfur dioxide in 1975, and predicts that figure may rise to 20 to 24 million tons by 1995. Most of the sulfur dioxide emissions originate in the heavily industrial upper Ohio Valley; from here, prevailing westerly winds carry them over eastern North America. Among the fossil fuels, coal is the worst culprit. U. S. coal reserves typically contain more than 3 percent sulfur, compared to only a few tenths of a percent for petroleum. For this reason, the EPA estimates that President's Carter's coal conversion plan could lead to a 10 to 15 percent increase in acid rain.

If coal-fired power plants are the largest single source of the sulfur dioxide component of acid rain, where do most of the nitrogen oxides (NO_X) come from? Gasoline-powered automobile engines contribute 40 percent of the total. In areas of extremely heavy traffic, such as Los Angeles County, nitrogen oxides from transportation contribute more heavily to acid rain than does sulfur dioxide. However, the next largest single source of nitrogen oxides is, once again, electrical power plants, which contribute 30 percent of the total. *This means that the generation of electrical power alone produces roughly half of the acid rain that falls across the continent.*

LAKES AND FORESTS UNDER SIEGE

In 170 mountain lakes in the Adirondacks, no fish swim—the waters are too acidic to support them. Two-thirds of the 1500 lakes in the million acre Boundary Waters Canoe Area on the Minnesota-Ontario border are approaching a pH of 5, which is critical for fish. Although the Great Lakes are big enough to dilute a lot of acid, two bays of Lake Huron are becoming dangerously acidic, and even in the mountains of western North Carolina, the pH of streams and lakes dips alarmingly after major storms, although for now they are still able to recover.

Mountainous areas such as the Appalachians and the Adirondacks are especially prone to receive acid rain, as moisture-laden air masses rise and cool over the mountains. Ironically, these areas are also the most sensitive to acid rain, for their thin soils are weathered from siliceous bedrocks such as granite, which have little capacity to buffer or neutralize the acid.

Aquatic ecosystems such as mountain lakes are particularly sensitive to changes in acidity. Healthy lake water may have a pH as high as 8 because of the presence of calcium bicarbonate. Acidification removes the calcium, and at pH 7 the declining calcium levels may affect the hatching of salamander eggs in the water. As the pH drops toward 6, snails and small crustaceans begin to disappear. The number and kinds of species that form the lake's complex food web decline rapidly, and the phytoplankton at its base begin to die. Bacterial decomposers are also killed, and organic matter builds up on the bottom.

Changes in the calcium balance disrupt ion exchange across the gills of fish, and also prevent fish egg production. Toxic metals such as mercury, released by the acidity, pose an additional threat. As the pH approaches 5, acid-loving mosses, fungi, and algae choke out the lake's other plants. More fish species die. Sphagnum moss creeps into the water, further depleting the calcium. At pH 4.5, all the fish and most of the frogs and insects are dead. The lake is clear and blue—and nearly lifeless.

The impact of acid rain on forest ecosystems is less well known because of the difficulty of separating the effects of numerous environmental pollutants. Growing evidence from laboratory and field studies, however, suggests that acid rain attacks the forest on two fronts. As it seeps into the soil, it leaches away valuable mineral nutrients while making toxic metals more soluble, so that plant roots take up these poisons instead of the lost nutrients. Soil microorganisms that break down organic matter and recycle its nutrients are killed by the acid, making the soil poorer still. Even the nitrogen-fixing bacteria that form nodules on the roots of plants such as legumes become scarcer as the pH drops.

Acid rain attacks the foliage as well as the soil, eating through the protective waxy cuticle of leaves, damaging photosynthetic tissue and distorting the chloroplasts. Trees thus weakened may be more susceptible to attack by disease and insects. Young trees are especially sensitive to acid rain, which inhibits seed germination, stunts seedling growth, and inhibits bud formation. The growth of entire forests in Scandinavia and the northeastern U. S. may be retarded by decreased photosynthesis and other effects. There is accumulating evidence that acid rain may also damage crop plants such as bush beans, soybeans, radishes, lettuce, tomatoes, and apples.

SOLUTIONS

What can be done to halt the blight of acid rain? The stop-gap measure of adding lime to lakes may keep them alive until some long-term solution is found, but it is an expensive and tricky business. Sweden spent $2.5 million on liming lakes in 1979, but 10 times that amount would be needed to treat all 20,000 of her dying lakes. Adding lime may actually increase the toxicity of metals released by the acid, so it has sometimes killed the trout and salmon it was designed to save. The New York State Department of Environmental Conservation has been trying the technique in a few lakes, but estimates that it costs $10 to $20 per pound of trout. The neutralizing effect of the lime may last several years—or only a few months.

Surely it is more reasonable to try to control the damaging emissions at their source. One option is to use low-sulfur fuels, but instead the U. S. now plans to adopt a $10 billion coal conversion plan which may substantially aggravate the problem.

Is it possible to burn coal cleanly? Washing coal before it is burned can reduce emissions by 10 percent, but it creates massive amounts of solid wastes to be disposed of. The most successful technique is flue-gas desulfurization, which uses solutions of limestone or other alkalis to absorb the sulfur dioxide from the flue gases. Such "scrubbers" are in operation on new power plants built to conform to the EPA's ruling that 70 to 90 percent of the sulfur dioxide must be removed from their emissions. Existing power plants, however, are not covered by this regulation, and this is why the pending coal conversion is expected to increase the fall of acid rain.

The first EPA ruling aimed directly at the problem of acid rain came in June of this year, when the agency ordered two Ohio power plants to reduce their sulfur dioxide emissions by 100,000 tons per year. The standards for these plants had been relaxed a year earlier because the Ohio coal industry complained that it would suffer if it had to purchase lower sulfur coal from out of state. A new computer model of the sources and movements of pollutants led the EPA to overrule the industry's protests. The Agency is also considering requiring control devices on older power plants, and the possible early retirement of the worst offenders among them.

All such control measures, however, cost money. How much should we be willing to spend when our data on acid rain are still sketchy, though suggestive? And how much *must* we know before we act? Groups such as the American Petroleum Institute and the Electric Power Research Institute argue that controls are unwarranted at present because existing data are too poor to establish beyond a doubt that acid rain is indeed due to human activity, or even that it is actually increasing, or that it damages the environment.

Numerous research programs are under way to try to find answers to the many questions that remain about acid rain: What is the mechanism of acid formation in the atmosphere? What will a more complete monitoring network reveal about increasing acidity and the spread of acid rain? Which regions of the continent are most sensitive? What roles do natural sources and long range transport of pollutants play? How does acid rain affect soils, forest growth, and crop yields? Is the response of the Adirondacks lakes typical, or do they represent an especially sensitive ecosystem? Just how resilient *is* the environment?

F. H. Bormann, whose work at the Hubbard Brook Experimental Forest in New Hampshire has contributed much of what is known about the effects of acid rain on entire ecosystems, has pointed out that forest ecosystems have a remarkable capacity to purify the rainwater that filters through them. Forests, in fact, serve as a buffer between man and his pollutants—but their resilience and buffering capacity are not unlimited.

Rather than cavalierly relying upon natural ecosystems to buffer us against our own poisonous wastes, we should perhaps consider how much abuse we can afford to heap upon this invaluable resource. As Bormann points out, forests provide an extraordinary array of benefits to man—and they do it on solar energy alone. There is what Bormann calls an "inseparable linkage" between the degradation of natural, solar-powered systems of benefit to man and the proliferation of fossil-fuel-powered human systems. Thus the more coal we burn for energy, the more our acid rain poisons the solar powered lakes and forests that produce food and lumber, filter our air and water, control erosion, and provide us with recreation and aesthetic pleasure. And the more we destroy these natural benefits, the more we must rely on polluting fossil fuels for energy to replace them. There are no easy solutions to this dilemma—only bitter ones.

Reprinted from *Carolina Tips*, Special Edition, September 1, 1980, Vol. 43, No. 9. by permission of Carolina Biological Supply Company, Burlington, North Carolina 27215, Gladstone, Oregon 97027

THE SOLID WASTE DILEMMA

INTRODUCTION: Suppose you went to the store to buy a few packs of gum. The clerk would hand them to you in a bag. After leaving the store, you might throw away the bag. Next, you open the pack of gum, throw away the outer wrapping, remove a single stick, and throw away the paper and foil wrapping around it. That's a lot of **trash** for one little piece of gum! Imagine how much **solid waste** you alone are responsible for each day. Now imagine how much your family produces. How about your school, community, state and country?

All over the world people are "throwing away" millions of tons of trash. But where is it going? Some solid waste, or **refuse**, may be **recycled**, some may be **incinerated**, and some may be put in **landfills**. Human and animal wastes may be chemically treated and turned into **sludge**.

A great deal of solid waste, however, is not adequately removed from the environment. Much of it can not be burned without producing dangerous by–products. New synthetic materials, like plastics, never decompose. **Toxic** materials cannot be left in land areas where they could contaminate the air or water. Our planet is rapidly running out of room for all this junk.

In this investigation you will determine the amount of trash you and your family produce each day. From this data, you can extrapolate the total weight of solid waste produced in this country.

PURPOSE:

- How much trash are you responsible for each day, and how can this solid waste be reduced or eliminated?

MATERIALS:

5 trash bags or large boxes
Marking pen
Bathroom scale

PROCEDURE:

Note: the majority of this activity will be done in your home

1. Obtain 5 large trash bags or cardboard boxes. Label them 1–5. Weigh each on your bathroom scale at home, and record its weight on Data Table 1.

2. Label the containers as follows:

    1 .... paper materials
    2 .... metal materials
    3 .... plastics
    4 .... glass and ceramics
    5 .... miscellaneous materials

3. Place the containers in a central area of your home, and ask the assistance of all your family members. Have everyone place their trash in the appropriately labeled container for the next two or three weeks. If possible, bring home any

trash that you, or other members of your family, produce while away from your home. **Do not include animal or plant materials (such as waste foods) in your collection.**

4. At the end of the collection period, reweigh each container, placing your measurements on Data Table 1. Discard (or recycle, if possible) the trash.

5. Calculate the total amount of trash produced by you and your family **per day** according to the following formula

$$\textbf{total waste per day} \quad = \quad \frac{\textbf{total weight of waste}}{\textbf{number of days}}$$

Enter this figure on Data Table 2.

6. Calculate the total amount of trash produced *per person*, per day, according to the following formula:

$$\textbf{waste per person/day} \quad = \quad \frac{\textbf{waste per day}}{\textbf{number of people in family}}$$

Enter this figure in Data Table 2.

7. Using a current reference source, determine the population of your community, state, and the country. Using the data from 6, above, extrapolate the total amount of solid waste produced by your community, state, and country in **one day**. Enter these figures on Data Table 2.

8. Using the figures you calculated above, determine the amount of solid waste produced in this country in **one year**. Enter this figure on Data Table 2.

DATA:

DATA TABLE 1: COLLECTION OF SOLID WASTE

Container #	Weight Empty	Weight Full	Weight of Trash
1			
2			
3			
4			
5			
		Total	

DATA TABLE 2: AMOUNT OF SOLID WASTE COLLECTED

Start Date	End Date	Total Days	Total Weight	Number of People	Weight per Day	Wgt. per person per day	Total Community.	Total State	Total U.S.A.	U.S.A. per Year

CONCLUSIONS:

1. What type of solid waste is produced by you and your family in the greatest amount by weight? The next greatest?

2. Based on your answer to question 1, do you see any relationship between the environmental problems of waste disposal and forest destruction?

3. Can you think of any way(s) to decrease the amount of waste produced by members of your family, particularly in the types of refuse listed in question 1?

4. Why is it important to decrease the amount of solid waste produced in this country?

5. Most solid waste is disposed of by either land disposal (dumps or landfill) or incineration. What are some benefits **and** drawbacks of these methods of waste disposal?

6. Recycling of waste materials, such as glass, metal and paper products, can greatly reduce the amount of solid waste that must be burned or buried. Why do you think that recycling is not as extensive as it could be?

SUGGESTIONS FOR FURTHER STUDY:

- Conduct an investigation, similar to this one, in your school. You will have to enlist the cooperation of the student body, faculty, administration and custodial staff, in order to collect significant data. Give careful thought to the way in which you collect and store the trash. Have your procedure approved by your teacher before you begin. After completion, write a report on your results and share it with the entire school. Involve your school government in possible ways to reduce the amount of solid waste being produced in your school and community.

- If your community has a landfill, research its operation. What are the proper conditions for successful management of a sanitary landfill? What state or local laws govern its operation? Are these laws or conditions being followed in your community? Is there any evidence of contamination by toxic materials? How long has the landfill been in operation? Does any burning take place at the landfill? How often is the waste covered by earth?

- If there is a waste incinerator in your area, investigate its operation in the same manner as described in the paragraph above.

- Aside from the mere volume of trash, why is today's solid waste a much greater problem than the wastes produced 30 years ago? What solutions, if any, do you see to this problem?

- Conduct a survey of the type of packaging used in both food and non–food items in your local supermarket. Redesign those packages that seem to include unnecessary "trash." Write to those companies that appear to be unconcerned with extra solid wastes, and demonstrate their excesses, requesting their cooperation in helping to control the ever increasing problem of solid waste disposal.

HOW AND WHERE DO BACTERIA GROW?

INTRODUCTION: Bacteria are unicellular organisms of the kingdom Monera. Since their genetic material is not enclosed in a nuclear membrane, bacteria are also known as **prokaryots**. There are many different species of bacteria, with a wide variety of nutritional and environmental requirements. Some bacteria cause disease, and are known as **pathogens**. Of the many nonpathogens, quite a few are beneficial, such as those used in the baking and brewing industries, bacteria of decay, and bacteria that provide nitrogen for plants.

Individual bacteria can only be seen with a high powered microscope. However, under proper environmental conditions, including a supply of food, bacteria can reproduce extremely rapidly. Once their numbers reach into the thousands, a growth pattern can be observed with the naked eye. On a solid **growth medium**, the result is called a **colony**. The specific shape and color of a colony can be used to help identify the bacteria that formed it.

In this activity, you will attempt to **culture** bacteria from various places, and investigate their different growth characteristics.

PURPOSE:

- Where can bacteria be found, and what is their pattern of growth?

MATERIALS:

Jar of nutrient broth and nutrient agar	Bunsen burner
4 sterile nutrient agar petri dishes	Wire inoculating loop
2 sterile nutrient broth tubes	Glass marking pencil
Sterile cotton swabs	Incubator
Hand lens	Masking tape
	Goggles

PROCEDURE:

Part I: CULTURING BACTERIA

While some bacteria require rather specific substances in order to grow, many can be cultured on common food sources, such as beef extracts, peptone and yeast extracts. Agar, a carbohydrate obtained from algae, may be added to the food extracts to solidify the medium. Agar itself is not a nutrient source. Nutrient agar is commercially prepared, and has appropriate food materials already added.

1. Look at a bottle of nutrient agar. What food source(s) does it contain?

Nutrient broth is another medium used for culturing bacteria. However, the agar is lacking. This liquid medium is sometimes preferable to a solid one, depending upon the purpose of the experiment.

2. Obtain a bottle of nutrient broth. How does its food content compare with the nutrient agar's?

In order to culture only the bacteria you actually want, the media and their containers must first be **sterilized**. In most classrooms, this is accomplished using an

autoclave, or pressure cooker. Your teacher will supply you with nutrient agar plates, and tubes of nutrient broth that have already been sterilized. **Do not open any of the dishes or tubes until you are ready to inoculate them.**

There are several methods of introducing bacteria into your culture medium. You will experiment with three of these.

I: Exposure Plates:

Exposure plates are used when bacteria in the air are to be cultured. By exposing the nutrient medium to the air, you allow bacteria to settle on the agar. Any bacteria present will multiply, and show up in the dish as colonies.

Obtain two sterilized nutrient agar petri dishes. Seal one with tape, and label it C, for control. Write your initials and today's date on the dish.

Select an area of your classroom or school where you think bacteria may be present in the air. Place your petri dish in this area and remove its top for a 15 minute period. (Do not place the top of your dish down on any surfaces). Seal your experimental dish and indicate the date and location of exposure on your data page. If necessary, draw a map showing where your dish was placed. Write your initials on the dish, and put both the experimental and control plates in an incubator at 30–35°C for 24 hours.

II: Contact Plates:

Suspected bacterial sources may be placed directly on the surface of the nutrient medium using a sterile cotton swab. Again, if present, the bacteria will form colonies on the surface of the agar.

Obtain two sterile nutrient agar plates and five sterile cotton swabs. For your control dish, gently rub the surface of agar with a newly opened swab. **Hold the top of the petri dish directly over the exposed plate while you inoculate it. Do not lift the cover any further than you have to.** (see illustration 1)

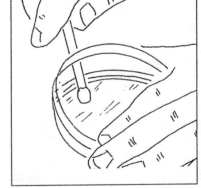

illus. 1

3. Why do you think these instructions are important?

Turn your other petri dish upside down without removing the lid. With a glass marking pencil, divide your dish into four quarters. Number the quadrants 1, 2, 3 and 4. (see illustration 2) Decide upon four areas you would like to test for bacterial presence. Any solid surface may be used, such as the floor, a lab counter, your hands, your teeth, or the inside of a piece of lab equipment. On your data page, write the locations from which you will inoculate each quadrant of your dish. Be specific.

For each of your sites, rub the surface with a newly opened sterile swab. As quickly as possible, rub the inoculated swab over the preselected quadrant of your dish. Remember, open the top of the dish only as much as needed to swab one quadrant. Quickly replace the lid while preparing for your next sample. Continue until all four quadrants have been contacted. Seal and initial the dish. Place both the control and experimental dishes in the incubator.

III: Wire Loop Inoculation:

When transferring bacteria, scientists and technicians often use an instrument called a wire inoculating loop. In order to introduce suspected bacteria into your broth tubes, you will employ this technique.

First select a possible source of bacteria. For this trial, choose a non–solid material, such as water (tap, pond, stagnant?), milk, yogurt, or cottage cheese. Write your choice on the data page.

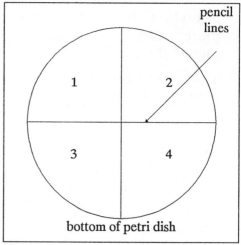

illus. 2

Bring your bacterial source, two sterile nutrient broth tubes, a wire loop and a Bunsen burner to your work station. **If you are not familiar with how to operate a Bunsen burner, consult your teacher before proceeding. Always wear goggles and tie back long hair when working with an open flame.** The wire loop will be used to transfer material from your selected source to the nutrient broth. The loop must be sterilized, however, both before and after contact with the source. This is accomplished by following the procedure below: (see illustration 3).

a) Light the bunsen burner and adjust the flame.

b) Pick up the broth tube in one other hand and keep it ready.

c) Pick up the inoculating loop in your other hand, and put the small, round end into the hot part of the bunsen burner flame until it glows red.

d) Immediately dip the sterilized tip of the loop into your source material.

e) Using the pinky and fourth finger of the hand holding the wire loop, remove the plug from the broth tube. Insert the tip of the loop, only, into the liquid.

f) Flame the mouth of the broth tube by passing the open end of the test tube through the hot part of the flame 2–3 times.

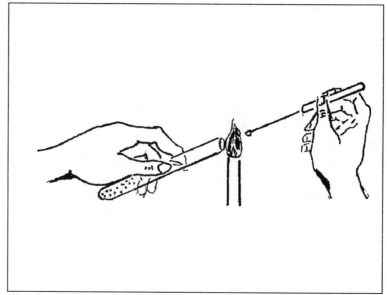

illus. 3

g) Replace the plug in the broth tube.

h) Flame the loop again.

Date and initial your tube.

Prepare a control tube by repeating the above procedure, but without using any source material for the inoculation. Label this tube appropriately.

Place both tubes in the incubator for 24 hours.

Part II: OBSERVING BACTERIAL GROWTH

On agar plates, different types of bacteria will grow in unique colonial forms. Observing the details of colony growth can aid in the identification of specific bacteria. In broth tubes bacteria also grow in different patterns depending upon the type. They do not, however, form recognizable colonies in the broth. You will now observe your tubes and plates, collecting data on bacterial growth. **Do not open any of the tubes or plates!**

Examine your exposure plates. Accurately draw their appearance in your Data section.

1. Are there any colonies present in the control dish? If so, how do you account for their presence?

2. Describe any growth you see in the experimental dish. Do you see individual colonies or total growth over the entire surface? What does this observation tell you about the numbers of bacteria that grew on the agar?

3. If you can see individual colonies, look at them under a steriomicroscope and study their characteristics. Do they all look the same, or do you have more than one type of bacteria growing on the dish?

4. Look closely at the edges, elevation, and form of any colonies you see. Refer to the reference chart in your Data section and classify each colony in the Data Table.

5. Repeat the above steps for your contact plate and control dish. Record your data and answer the observation questions. (5–8 parallels 1 - 4)

9. Observe your nutrient broth tubes. Describe the color and clarity (turbidity) of the control tube. Is there any evidence of bacterial growth in this tube? If yes, how do you account for this?

10. Referring to the reference chart on surface growth in nutrient broth, classify and record any growth you observe in your experimental tube.

DATA:

Part I:

1._____

2._____

Location of experimental exposure plate: _____

3._____

Location of contact plates:

Quadrant 1:_____   Quadrant 3:_____

Quadrant 2:_____   Quadrant 4:_____

Bacterial source for nutrient broth tube: _____

Part II:

Control Exposure Plate   **Experimental Exposure Plate**

1._____

_____

2._____

_____

3._____

_____

REFERENCE CHART

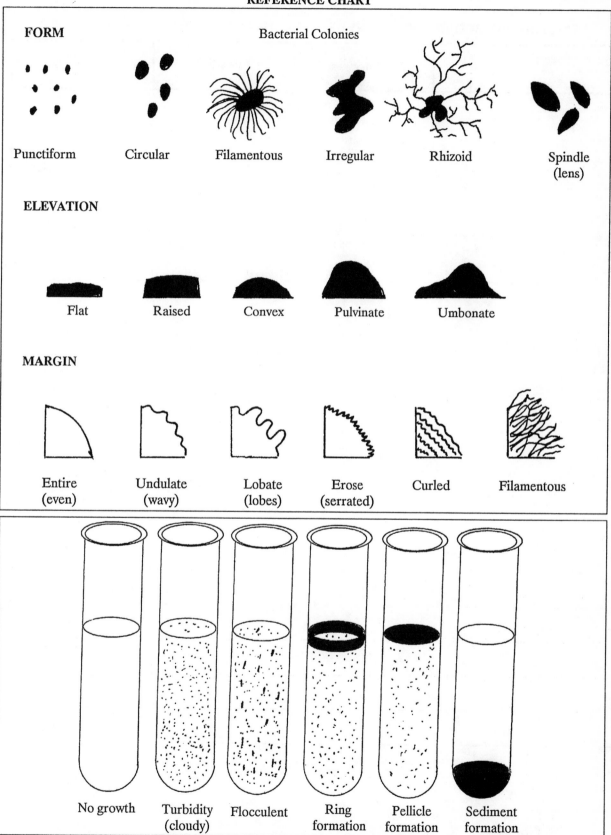

FORM

Bacterial Colonies

Punctiform   Circular   Filamentous   Irregular   Rhizoid   Spindle (lens)

ELEVATION

Flat   Raised   Convex   Pulvinate   Umbonate

MARGIN

Entire (even)   Undulate (wavy)   Lobate (lobes)   Erose (serrated)   Curled   Filamentous

No growth   Turbidity (cloudy)   Flocculent   Ring formation   Pellicle formation   Sediment formation

CHARACTERISTIC GROWTH PATTERN: NUTRIENT BROTH TUBES REFERENCE

4.

EXPERIMENTAL RESULTS OF COLONY GROWTH CHARACTERISTICS–EXPOSURE PLATE

Colony	Form	Edges	Elevation	Color
1				
2				
3				
4				

Control Contact Plate

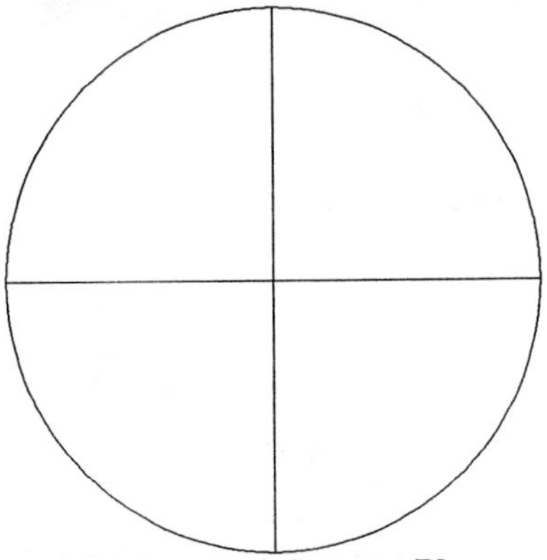

Experimental Contact Plate

6._____

_____

7._____

_____

8._____

_____

9.

EXPERIMENTAL RESULTS OF COLONY GROWTH–CONTACT PLATE

Colony	Form	Edges	Elevation	Color
1				
2				
3				
4				

10._____

_____

11. Results of your broth tube :

CONCLUSIONS:

1. What was the purpose of the control tubes you set up in Part I?

2. Why must all bacteriological equipment and materials be sterilized before use?

3. How did the types and numbers of colonies seen on the exposure plate compare with those of the contact plate?

4. Give three possible reasons for any apparent lack of bacterial growth in any of the experimental set–ups.

SUGGESTIONS FOR FURTHER STUDY:

- If you have access to a colony counter, use it to determine the number of colonies grown on each of your experimental plates. What conclusions can you draw from these data?

- Conduct a comparative study of bacterial presence in various water sources. Describe your experimental design and accurately record your results. What do your results indicate?

- Ask your teacher for instructions on the preparation of a bacterial smear. Prepare slides from the different colonies you obtained on your experimental dishes. Draw and classify the various bacteria you find.

WHAT DO BACTERIAL ENZYMES DO?

INTRODUCTION: Bacteria, although single–celled, are complete living organisms. As such, they must carry out all the metabolic activities associated with life. In order to accomplish these activities, bacteria, like other living systems, must employ **enzymes**, or organic catalysts. Some of these enzymes are **hydrolytic**, converting the complex substances in the environment into simpler molecules that may diffuse into the bacteria cell. Other enzymes may catalyze **synthesis** reactions, providing the bacterial cell with usable compounds. Still other **metabolic** enzymes are needed to convert toxic wastes into harmless compounds, convert food energy into ATP through respiratory processes, and perform many other essential biochemical functions.

In this exploration you will investigate the actions of several bacterial enzymes, including those that hydrolyze starch and protein, and those that convert hydrogen peroxide and nitrates into other compounds.

PURPOSE:

- What are the functions of various bacterial enzymes?

MATERIALS:

Bacterial cultures: E coli,
  B. subtlis, B. megatherium
Sterile petri dishes:
  1 starch agar, 1 nutrient agar
3 sterile gelatin agar tubes
3 sterile nitrate broth tubes
Hydrogen peroxide
NI and NII solutions
Wire loop
Bunsen burner

Lugol's iodine solution
Benedict's solution
Boiling water bath
Goggles
Incubator
6 test tubes
Zinc (a pinch)
Hydrogen peroxide
Teasing needle

PROCEDURE:

Part I: STARCH ENZYMES

In order for bacteria to utilize starch as a source of food, they must be able to hydrolyze it into smaller, simpler sugars. This reaction requires an enzyme, **amylase**, that is capable of catalyzing this hydrolysis.

1. Obtain a sterile petri dish containing starch agar.

2. Streak one side of a starch agar petri dish with the bacteria E. coli. Streak the other side of the dish with B. subtlis. (See illustration 1). If you have not done investigation B43, *How and Where do Bacteria Grow,* follow the directions below for performing a bacterial transfer. If you are already familiar with this procedure, go directly to step 3.

How To Transfer Bacteria

Bring a wire loop, a Bunsen burner and the bacterial culture tubes to your work station. **If you are not familiar with how to operate a Bunsen burner, consult your teacher before proceeding. Always wear goggles and tie back long hair when working with an open flame.** The wire loop will be used to transfer material from the culture tube to the petri dish. The loop must be sterilized, however, both before and after contact with the source. This is accomplished by following the procedure below:

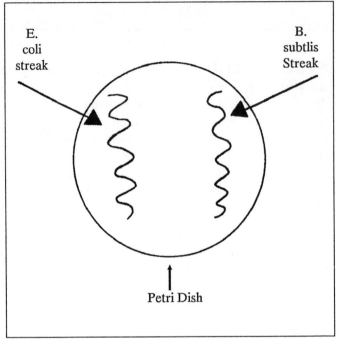

illus. 1

a)   Light the bunsen burner and adjust the flame.

b)   Pick up the culture tube in one hand and keep it ready.

c)   Pick up the inoculating loop in your other hand.

d)   Using the pinky and fourth finger of the hand holding the wire loop, remove the plug from the culture tube.

e)   Flame the mouth of the culture tube by passing it through the flame of the Bunsen burner two or three times.

f)   Flame the tip of the wire loop by holding it in the hot part of the burner flame until it glows red.

g)   Quickly slide the tip of the loop across the surface of the culture so that you hear a "sizzling" noise.

h)   Reflame the mouth of the culture tube and replace the plug.

3.   Streak the bacteria onto the petri dish by touching the tip of the wire loop, containing the inoculum, onto the surface of the agar, and spread it, in a serpentine motion, across the surface. **Do not puncture the surface of the agar.** Remember to keep each bacteria on its own side of the petri dish.

4.   Label the plate with your initials and the date. Incubate the petri dish at 35°C for 24 hours.

Proceed To Part II: Return To Step 5 After 24 Hours

5.   Remove the cover from your petri dish, and flood the dish with Lugol's iodine solution. If the area around either bacterial streak turns a blue–black color, starch is still present, and has not been hydrolyzed. If the iodine remains its

original color, no starch is present, indicating that the bacteria have converted the starch into another substance, presumably sugar.

6. To determine if the new substance is indeed sugar, remove a small sample of material that showed no starch, and add it to a test tube with some water and Benedict's solution. Shake the tube to mix the contents. Heat the tube in a boiling water bath for three minutes to test for the presence of sugar. If the Benedict's solution turns green, yellow, orange, or brick red, sugar is present (in increasing quantities).

7. Record your data on Data Table 1.

Part II: HYDROGEN PEROXIDE ENZYME

Hydrogen peroxide, H_2O_2, is a toxic waste product produced by cells. It is broken down into the harmless products of water and oxygen by the enzyme **catalase**. The presence of catalase may be determined by bubbling activity when hydrogen peroxide is placed in contact with cells containing the enzyme.

1. Obtain a nutrient agar petri dish, and culture tubes of E. coli and B. subtlis. Prepare a streak plate as you did in Part I. Label the plate with your initials and the date. Incubate the plate for 24 hours.

Proceed To Part III: Return To Step 2 After 24 Hours

2. Remove the cover of the petri dish. Add several drops of hydrogen peroxide to the streaked areas of the dish. If bubbling occurs, the presence of catalase is confirmed.

3. Enter your results on Data Table 2.

Part III: PROTEIN ENZYME

Gelatin agar, a protein–enriched medium, is a solid at room temperatures. If the protein is hydrolyzed to amino acids, the medium liquifies. Bacteria containing **protease** enzymes, therefore, should liquify the medium.

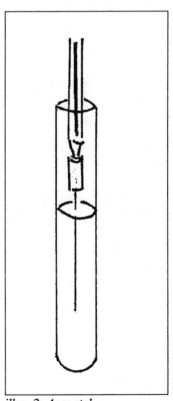

1. Using a straight teasing needle, inoculate each of 3 gelatin tubes with E. coli, B. subtlis and B. megatherium. The inoculation should be done using a technique known as an agar stab. (See illustration 2). First flame a teasing needle and the culture tube. Next, insert the needle directly into the culture medium, and withdraw it slowly, being careful not to touch the sides of the tube. Reflame the culture tube and plug. Insert the inoculated needle into the gelatin, and withdraw carefully. Flame the needle. Flame and cap the gelatin tube.

2. Allow the tubes to stand at room temperature for 24 hours.

illus. 2: Agar stab

Proceed To Part IV. Return To Step 3 After 24 Hours

3. Observe the gelatin tubes for any signs of liquification.

4. Record your data on Data Table 3.

Part IV: NITRATE ENZYME

Some bacteria have the ability to combine hydrogen (given off in many biochemical reactions) with nitrates to produce nitrites and water:

$$H_2 + NaNO_3 \longrightarrow NaNO_2 + H_2O$$

This reaction is important to the recycling of nitrogen compounds in the environment. This reduction reaction can be identified in the laboratory, and thus the presence of the necessary enzymes may be determined.

1. Obtain 3 nitrate broth tubes, and culture tubes of the bacteria E. coli, B. megatherium and B. subtlis.

2. Inoculate each of the nitrate tubes with one of the bacteria cultures.

3. Label each tube with your initials, the bacteria present, and the date.

4. Plug with sterile cotton and incubate for 24 hours.

5. Divide each of your broth tubes in half, labeling accordingly. Perform the following test on each pair of tubes.

6. Add 2 drops of NI and 2 drops of NII to one of the two tubes. If a red or pink color appears, the organism changed (reduced) the nitrate to nitrite.

7. If the first tube was negative (no color change), add a pinch of zinc to the second tube. After 4 minutes, add 2 drops of NI and 2 drops of NII. If a pink color appears now, it confirms that the organism does **not** contain the necessary enzyme, and is therefore a non–reducer.

8. Record your results on Data Table 4.

DATA:

Part I

DATA TABLE 1: AMYLASE ACTIVITY IN BACTERIA

Nutrient present After 24 Hrs.	E. coli	B. subtlis
STARCH		
SUGAR		

DATA TABLE 2: CATALASE ACTIVITY IN BACTERIA

	E. coli	B. subtlis
H_2O_2 Activity		

DATA TABLE 3: PROTEASE ACTIVITY IN BACTERIA

	E. coli	B. subtlis	B. megatherium
Evidence of Liquification			

DATA TABLE 4: NITRATE ENZYME ACTIVITY IN BACTERIA

	E. coli	B. subtlis	B. megatherium
N I			
N II			

CONCLUSIONS:

1. Summarize your data by indicating the enzymes contained in the following bacteria:

   E. coli _____

   B. subtlis _____

   B. megatherium _____

2. According to your answers above, what processes are possible, if any, in E. coli but NOT in B. subtlis? In B. subtlis but NOT E. coli?

3. How can information about enzymes be used to help identify bacteria?

SUGGESTIONS FOR FURTHER STUDY:

- Bacteria use sugars as a source of energy. They can oxidize the sugar during the process of respiration, produce ATP, and give off waste products. One of these wastes is CO_2. Enzymes are essential to bacterial respiration, as they are to respiration in any organism. Can one bacteria, however, utilize *any* sugar for respiration? Does it have enzymes that will act on any sugar, or are they specific for only one or two types of sugar? This question can be investigated by measuring the amount of waste gas (carbon dioxide) given off by bacteria when they are supplied with a variety of sugars. If no gas is released, the bacteria can not utilize that particular sugar, and therefore do not possess the proper enzymes. The more gas produced, the more easily the bacteria can oxidize the sugar.

  Design and conduct an experiment to answer the following question: "Do bacteria contain enzymes that will oxidize different types of sugars?" Some possible sugars to use include glucose, sucrose, lactose, maltose, fructose, galactose. E. coli is a common bacteria that may be used for this experiment, or your teacher may suggest another source.

  Fermentation tubes should be used to conduct your experiment. If gas is produced, it will displace the sugar solution in the tube, lowering its level in the tube's arm. The amount of gas can be determined by measuring the difference in broth level from the start to the end of the experiment.

  Be sure to have your experiment approved by your teacher before proceeding. Report your results.

HOW PURE IS YOUR WATER?

INTRODUCTION: Water is one of our planet's most precious resources. Fresh water ponds, streams and lakes are home to a wide variety of organisms. Billions of people depend upon clean, fresh water for drinking. Rivers and oceans must be kept clean and safe for plant and animal life. Increased world population and industrialization, however, have led to increasing threats to the purity of our water supply. Both biological and industrial **pollutants** are increasingly contaminating our water. In order to minimize the destruction of pure water sources, and reverse the trend of pollution, careful monitoring and analysis of our water supplies are required. In this activity, you will perform some simple diagnostic tests to determine the "purity" of your local water.

PURPOSE:

*   What is the quality of your water?

MATERIALS:

2 (or more) water samples	Goggles
2 collecting flasks: 250 ml	Filter paper
Rubber stoppers for above flasks	Funnel
Sterile pipettes	50 ml beaker
Dissolved Oxygen Test Kit	Microscope slides
Test tubes	Compound microscope
Stoppers for tubes	Colorimeter (optional)
pH paper	Other water test kits (optional)
250 ml flask	

PROCEDURE:

Collect water samples from at least two sources. One of these sources should be drinking water from home or school. Another should be from a local fresh water source, such as a pond or stream. Your teacher will tell you to collect other samples, if desirable, as well as instruct you in any safety procedures to follow. A water sample should be collected by filling a flask (at least 250 ml) to overflowing, then securely closing it with a rubber stopper. Label the flask with the location and date of the sample. Be sure to rinse the collection flask several times with the sample water. When collecting water from a pond or stream, take your sample from *below* the surface. When collecting drinking water, let the water run for several minutes before taking your sample.

Perform the following tests on both samples. Record your observations on Data Table 1:

COLOR: Observe each of your water samples against a white background and note any apparent color. Greater accuracy in assessing color in your water samples may be achieved by filling a flask with distilled water and holding it next to your sample for comparison. Colors may include colorless, brownish, greenish or "other." If a colorimeter is available, ask your teacher for instructions in its use.

TURBIDITY: The appearance of your water sample may be clear, hazy or cloudy, due to the presence of fine particles suspended in the water. Turbidity can be assessed in the same manner as color.

ODOR: Cautiously sniff each of your samples. Describe the smell, if any. Examples of possible odors include musty, chemical, rotten eggs, none, etc.

DETERGENTS: The presence of detergents in the sample may be observed in the following manner: Pour about 20 ml of each sample into separate test tubes and stopper securely. Shake the tubes vigorously for one or two minutes. Look for foam on the surface of the water. If it is present, detergent is in the water sample.

OIL: Examine your sample's surface for the presence of oil film or scum.

pH: Using a sterile pipette, remove a drop of water from each sample and place it on a strip of pH paper. Compare the color of the indicator paper with the color chart provided, and record the pH of each sample. Be sure to use a separate pipette and piece of pH paper for each sample.

DISSOLVED OXYGEN (DO): The amount of oxygen dissolved in water is critical for the survival of plant and animal life. Oxygen is added to water by direct absorption from the air and by the action of photosynthetic plants. During respiration and decomposition, oxygen is removed from the water. The amount of oxygen dissolved in any water sample decreases as the temperature of the water increases. Salt water can hold less oxygen than fresh water. The minimum amount of dissolved oxygen that can support aquatic life safely varies between 4.0–5.0 parts per million (mg/liter). The maximum that can be expected at room temperature is approximately 8.0 ppm for fresh water, and 7.0 ppm for salt water.

If a DO water test kit is available, use this to determine the dissolved oxygen in each of your samples. If such a kit is NOT available, the DO may be determined by following the instructions below:

1. Carefully pour 100 ml of your sample into a clean flask. Avoid bubbling or splashing when you transfer your sample, since this will add oxygen to the water.

2. Add 10 drops of manganese sulfate solution, and 10 drops of alkaline potassium iodide solution to your sample. Swirl the flask gently to mix the solutions.

3. Let the flask stand for 2 minutes (10 minutes for salt water).

4. Add 15 drops of concentrated sulfuric acid. **Wear goggles when using this acid. Use extreme caution when handling concentrated acids.** Swirl gently to mix.

5. While continuing to swirl the flask, add starch solution, one drop at a time, until a blue, or blue–green color appears and remains.

6. Add sodium thiosulfate solution, one drop at a time, until the solution in the flask becomes colorless. You must *count the number of drops* of thiosulfate added. When the solution begins to decolorize, swirl the flask gently to make sure the colorless condition is permanent. Record the total

number of drops of thiosulfate that is needed to permanently decolorize the sample.

7. To convert the number of drops of thiosulfate into ppm of dissolved oxygen, simply divide the number of drops by 30. Record the dissolved oxygen content, in parts per million, on the Data Table.

FILTRATE: Filter 25 ml of sample through filter paper that has been folded into a funnel. Be sure to stir your samples before filtering them. Observe and record the appearance of the filter paper and the filtrate.

MICROORGANISMS: Prepare several wet mounts of each water sample. Take samples from the top, middle and bottom of each flask. Observe your slides through a compound microscope, and sketch any organisms found in each slide. Use a separate piece of paper for your sketches. Be sure to indicate the sample used, and depth of sample on each sketch. Summarize your microscopic findings on the Data Table.

OTHER TESTS: If your teacher has other water quality test kits available, use them to determine other characteristics of your water, as instructed. Add this data to the Data Table.

DATA:

DATA TABLE 1: WATER SAMPLE CHARACTERISTICS

Condition	Drinking Water	Fresh Water	Other (if used)
COLOR			
TURBIDITY			
ODOR			
DETERGENTS			
OIL			
pH			
DISSOLVED OXYGEN			
FILTRATE & PAPER			
MICROORGANISMS			

CONCLUSIONS:

1. What do your results from the tests for color, turbidity, odor, oil and detergent imply about the quality of your drinking and fresh water?

2. Most natural waters have a pH range between 5.0–8.5. Freshly fallen rain should have a slightly acidic pH, about 5.5–6.0. Water that reacts with soils and minerals containing alkaline materials may have a pH of 8.0–8.5. Sea water has an average pH of 8.1. What do your results from the pH tests tell you about the quality of your water? Is there any evidence of an "acid rain" problem in your water?

3. What do your results from the filtration and microscopic studies show about the quality of your water?

4. Why is the amount of dissolved oxygen in water an important piece of information? What do your results imply about your fresh water sample? Why would the DO content decrease in polluted waters?

5. Why must the quality of drinking water be continually tested?

SUGGESTIONS FOR FURTHER STUDY:

- The presence of coliform bacteria is often used to test for contamination of fresh water sources. These bacteria are commonly found in the intestinal tract of animals, including humans. Where sewage treatment is lacking or inadequate, these bacteria can find their way into the drinking water. The presence of these contaminants indicate that other, more dangerous pathogens, like typhoid, may also have gotten into the water. Coliform bacteria can be detected by a rather simple lab test, since they are one of only a very few bacteria that can metabolize lactose broth and produce gas. Collect samples of water from various local sources. Test these for the presence of coliform bacteria. If your teacher has a coliform test kit, use it. If such a kit is not available, follow the procedures outlined below. **Use caution when working with samples that may be contaminated. Maintain aseptic technique throughout the procedure.**

  1. Obtain a fermentation tube of lauryl tryptose broth for each sample.

  2. Open one tube, and flame the mouth of the tube in a Bunsen burner. **Wear goggles.**

  3. Using a sterile pipette, transfer 1 ml of the water to be tested into the broth.

  4. Reflame the mouth of the tube and replace the cap.

  5. Invert the broth tube so that the small, inner fermentation tube fills with broth.

  6. Turn the tube in an upright position and incubate at 35°C for 24 to 48 hours.

  7. Repeat the inoculation procedure for each of your water samples.

  8. Examine the fermentation tubes for the presence of gas. This indicates the probable presence of coliform bacteria. The more gas present, the more bacteria are in the sample.

  9. Further tests may be done to confirm their presence. Ask your teacher if such tests are possible. If so, get instructions and perform these tests.

  Write a report on your experiment and its results.

WHAT KILLS GERMS?

INTRODUCTION: You have probably seen many ads on T.V. or in a magazine in which a certain product claims to "kill germs." But what is a germ? Biologically, there is no such thing! When non–scientists mention "germs," they are usually referring to organisms that cause disease, or **pathogens**. Most of these pathogens (but not all) are among a group of single–celled organisms commonly called **bacteria**, belonging to the kingdom **Monera**.

Bacteria can multiply extremely rapidly. It is often desirable or necessary to stop their growth, whether it is in your body, on your food, or in your kitchen. To do this, products such as **disinfectants**, **antiseptics**, and **antibiotics** are used. Antibiotics differ from antiseptics and disinfectants in that they are produced by living organisms (usually plants or molds). Some of these products actually kill the bacteria (bacteriocidal), while others merely stop their continued growth (bacteriostatic). In this exploration you will investigate the effectiveness of some of these products.

MATERIALS:

Cultures of: B. subtlis, E. coli
6 antibiotic disks: 2 each
  of 3 different types
3 different disinfectants/antiseptics
10 blank, sterile disks
4 sterile nutrient agar petri dishes
4 sterile cotton swabs
Bunsen burner

Goggles
Sterile forceps (in alcohol)
Incubator
Metric ruler
Stereomicroscope or
  hand lens (optional)
Glass marking pencil

PROCEDURE:

1. Obtain 4 sterile nutrient agar petric dishes. Bring the culture tubes of B. subtlis and E. coli to your work station, along with a Bunsen burner and 4 sterile cotton swabs. **Always wear goggles when working with the Bunsen burner.**

   If you have not done explorations 43 or 44, follow the directions, given at the conclusion of the procedure section, for transferring bacteria and making a streak plate. Be sure to read all directions before beginning.

2. Thoroughly streak two of the petri dishes with B. subtlis, and the other two with E. coli. Be sure to completely cover the surface of the petri dish with the bacteria. Label each dish with the name of the bacteria used.

3. Label one of the E. coli dishes "A," and one of the B. subtlis dishes "B." Turn the dishes over and use a marking pencil to divide each

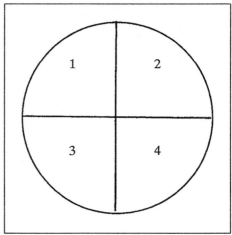

illus. 1

dish into four quadrants. Label the quadrants 1, 2, 3 and 4 (see illustration 1).

4. Select one of the available antibiotic disks, and using sterile forceps, place it in the center of quadrant 1 in dish A. (Try to open the petri dish as little as possible when you insert the disk). Press the disk down with the tip of the forceps so that it rests on the surface of the agar. Immediately replace the cover on the dish. Place a disk of the same antibiotic in quadrant 1 of petri dish B. In Data Table 1, record the antibiotic used.

5. Using two other antibiotics, repeat the above procedure for quadrants 2 and 3 of both dishes. Record the antibiotics used in each quadrant on the Data Table.

6. In quadrant 4 of both dishes, place a blank, sterile disk. Why is a blank disk used?

7. Divide the other two dishes of E. coli and B. subtlis in the same manner as before. Label these "C" and "D," respectively.

8. Into quadrants 1–3, you are going to place disks that have been dipped into 3 different antiseptics/disinfectants. To do this, remove a blank disk from its container with sterile forceps. Immediately dip it into one of your selected test solutions, shake off excess liquid, and place the disk in one of the quadrants of your marked petri dish, as before. Quickly cover the dish to avoid contamination. Record the substances used in each of the 3 quadrants on the Data Table.

9. Into quadrant 4 of both dishes, place a blank disk.

10. Label each petri dish with your initials and place it in an incubator at 37°C for 24–48 hours.

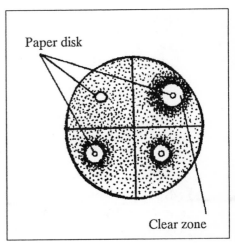

illus. 2

11. Remove the dishes and inspect them, one at a time, for bacterial growth. Look carefully at the areas immediately surrounding each of your disks. If there is a halo, or **zone of inhibition** around the disk, it means that bacteria were prevented from growing in the immediate vicinity of the disk. (See illustration 2). If a halo is present, measure its size, from the edge of the paper circle to the nearest bacterial growth, with a mm ruler. You may want to use a hand lens or stereomicroscope to make this measurement easier. Record all your observations in the Data Table.

12. Look closely in the zones of inhibition on all your dishes. Using a hand lens or stereomicroscope if needed, determine if any small, distinct **colonies** are growing in this area. Record the presence and description of any of these colonies on Data Table 2.

13. Return all your dishes to your teacher for proper sterilization/disposal.

HOW TO TRANSFER BACTERIA

Bring a sterile cotton swab, a Bunsen burner, a petri dish, and the bacterial culture tubes to your work station. **If you are not familiar with how to operate a Bunsen burner, consult your teacher before proceeding. Always wear goggles and tie back long hair when working with an open flame.** The cotton swab will be used to transfer material from the culture tube to the petri dish. When opening the culture tube, however, you want to avoid contamination as much as possible. This is accomplished in the following manner:

a) Light the Bunsen burner and adjust the flame.

b) Pick up the culture tube in one hand and keep it ready.

c) Pick up the sterile cotton swab (still in its protective wrapping) in your other hand. Rip open the top of the swab wrapping, but do not remove it yet.

d) Using the pinky and forth finger of the hand holding the swab, remove the plug from the culture tube.

e) Flame the mouth of the culture tube by passing it through the flame of the Bunsen burner two or three times.

f) Remove the swab from its wrapper, and quickly place it on the surface of the culture. Roll it over the surface so that bacteria will adhere to the entire tip of cotton. Do not jab the swab into the culture medium. Remove the swab.

g) Reflame the mouth of the culture tube and replace the plug.

PREPARING A STREAK PLATE

a) Place the petri dish on the table, and raise the lid *slightly*. You do not want to raise the lid more than about 45°. Working quickly, touch the cotton tip of your inoculated swab onto the agar at the top left corner of the petri dish.

b) Spread the bacteria on the agar by moving the swab tip back and forth across and down the petri dish. Do not press down too hard into the agar. You only want the bacteria on the surface, not inside the agar gel.

c) When you have completely covered the petri dish, replace the lid and discard the swab according to the directions given by your teacher.

DATA TABLE 1: RESULTS OF ANTIBIOSIS ON BACTERIAL GROWTH

Dish	Bacteria	Substance On Disk	Quadrant	Observations	Size of Zone of Inhibition (mm)
A	E. coli		1		
			2		
			3		
		none	4		
B	B. subtlis		1		
			2		
			3		
		none	4		
C	E. coli		1		
			2		
			3		
		none	4		
D	B. subtlis		1		
			2		
			3		
		none	4		

6. _____

DATA TABLE 2: GROWTH IN ZONES OF INHIBITION

Dish	A	B	C	D
Quadrant #				
Descripton of Colony				

CONCLUSIONS:

1. Different bacteria usually produce colonies with different characteristics. Is there any evidence that your petri dishes contain any bacteria other than the type you put there? Explain.

2. Describe the effects of the various antibiotics on the two types of bacteria you used. Were they all equally effective on both types of bacteria? Explain.

3. How might these antibiotic disks be used in medical research or diagnosis?

4. Which of the two bacterial strains is more sensitive to antibiotics? Explain.

5. How effective were each of the antiseptics you used? Would you use them to halt the growth of bacteria in your home or on your body?

6. Did the effectiveness of the antiseptics vary with the type of bacteria? Explain.

7. Is there any observable difference between the effectiveness of antiseptics and antibiotics? Explain.

8. How can you account for any colonies that grew within the zone of inhibition?

SUGGESTIONS FOR FURTHER STUDY:

- Common molds, such as Penicillium, have been shown to inhibit bacterial growth. As a matter of fact, many antibiotics are produced by molds. Penicillium, and other molds, may be grown on fruit, such as oranges. Try growing some mold on various food objects. Isolate a pure section of the mold and use it as you did the antibiotic disks in this investigation. Describe the mold you grew. Try to identify your molds by microscopic examination and the use of reference books. Report on their antibacterial qualities. Do your results reflect those of known drugs?

- Certain organisms (bacteria, fungi, mold) found in soil have the ability to inhibit bacterial growth. Try an experiment to see if organisms in your local soil have the ability to inhibit the growth of bacteria. Below is a suggested procedure.

   Collect a small sample of soil, from a depth of 10–20 cm, and suspend it in 100 ml of sterile distilled water. Remove 1 ml of the suspension (before it settles) and add it, with a sterile pipette, to a tube of liquified agar. Rotate the agar tube between your palms in order to thoroughly mix the contents. Pour the agar into a sterile petri dish and allow it to solidify. Incubate the dish at 35°C until you see the growth of colonies.

   Using a wire loop, isolate a sample of each different, distinct colony you find. Insert each inoculated loop into a separate sterile broth tube, and incubate it for 24 hours. On a data sheet, describe each of your colonies and give them an identification code. Label each broth tube with the identification code for its colony. You now have pure cultures of various microorganisms in the soil sample.

   Use each of your pure soil sample cultures as possible antibacterial agents. Do this by applying one streak of each culture to an agar plate that has been completely streaked with a bacteria that you select (E. coli is a good choice). After 24–48 hours, look for any evidence of bacterial inhibition. Report your results. Try to identify any of the microorganisms that inhibited bacterial growth.

- Write a research report on antibiotics. Some of the questions you might address include: What are the various types? What is the history of their discovery? How are they produced today? What is meant by broad spectrum drugs? How do sulfa drugs relate to antibiotics? What is meant by resistance to antibiotics? Be sure to include a complete bibliography with your report.

```
╔══════════════════════════════════════════════════════════╗
║                    TRIVIA CHALLENGE                        ║
╚══════════════════════════════════════════════════════════╝
```

INTRODUCTION: This activity is a game of recall designed to help you review a specific topic in your course of study. By preparing your own questions and answering those prepared by another student, you will review the details of any unit chosen. Use this activity to help you prepare for unit, mid–term, or final exams.

PURPOSE:

- To review a unit of study by writing questions and answering those questions written by other students.

MATERIALS:

3 x 5 index cards

PROCEDURE:

Determine the area of study for which you will prepare questions and play Trivia Challenge. Print your questions on one side of a 3 x 5 index card, one to a card, and include the answer on the reverse side. Prepare 10 to 20 cards for each topic. Your teacher will tell you the exact number of questions to prepare. Select an opponent from your classmates (or one may be assigned by your teacher) and sit facing each other. One of you will show your opponent a question by flashing your index card for 5 seconds. The responding student then has another 5 seconds to answer. If the response is correct, your opponent gets one point. If incorrect, you receive the point. Now switch roles so that your opponent flashes the question at you. Again, you may see the question for 5 seconds, and have another 5 seconds to answer. Record the points as before. Keep a running total of your scores until all the cards have been played. The player with the most points at the end of a round is the winner. NOTE: a bonus of 5 points will be awarded to any player who detects a mistake in his/her opponent's answers!

The following list is a sample of the types of questions that may be appropriate for Trivia Challenge:

Topic	Front of Card (Question)	Back of Card (Answer)
Skeletal system	Thigh bone	Femur
	Producer of growth hormone	Pituitary gland
	Jointed sections of backbone	Vertebrae
Ecology	Largest biomass in a food chain	Producers
	Stable animal and plant life in a biome	Climax community

SUGGESTIONS FOR FURTHER STUDY:

- Modify the types of questions you ask in Trivia Challenge to include diagrams and calculations.

- Organize your classmates into teams for a larger game.

- Organize your class teams to compete against teams from another class. Set up a Trivia Tournament.

JEOPARDY

INTRODUCTION: You are about to play a classroom version of the television game show Jeopardy. All the "answers" will be in biology topics related to your classwork. You will supply the correct "questions" and accumulate points.

PURPOSE:

- To accumulate the most points by correctly supplying questions in various biology categories.

MATERIALS:

Jeopardy "answers"

PROCEDURE:

Your teacher will help you get organized in order to play the game. As in the television show, there will be five related (biology) categories, with five "answers" of varying point values in each. More difficult questions have higher point values. Once a category and a point value has been selected, someone will read the answer aloud. As soon as you think you know the correct "question," raise your hand. If you are called on and your question is correct, you get the point value of that answer.

For example, if the teacher reads the answer:

"Contains chloroplasts and a cell wall"

The appropriate question would be:

"What is a plant cell?"

If your question is incorrect, the points will be deducted from your total.

If you play a second game, double jeopardy, there are five new categories, but with the point values doubled. Total points from both single and double jeopardy determine a winner.

SUGGESTIONS FOR FURTHER STUDY:

- Make up your own jeopardy game. Select a major topic (such as plants, classification, reproduction, etc.) and prepare "answers" in five related topics. Exchange your game with other students. Use these for review purposes.

GAME BOARD LAYOUT

BIOLOGY JEOPARDY				
Category 1	Category 2	Category 3	Category 4	Category 5
100	100	100	100	100
200	200	200	200	200
300	300	300	300	300
400	400	400	400	400
500	500	500	500	500

SAMPLE GAME
ANIMAL MAINTENANCE: JEOPARDY

Nutrition	Transport	Excretion	Respiration	Regulation
Type of organism unable to synthesize its own food from raw materials	Movement of materials across a membrane from low to high concentrations	CO_2, H_2O and nitrogenous compounds	$C_6H_{12}O_6 + 6O_2 \rightarrow 6H_2O + 6CO_2 + ATP$	Name for a nerve cell
Enzymatic hydrolysis of large molecules into smaller ones	Organism with an open circulatory system	Organ of gas excretion in the earthworm	This must be thin and moist	Organism possessing a nerve net
Organism that carries on both intra and extra cellular digestion	Hemoglobin	Nephridia is to earthworm as ------ is to human	Insect's organ of gas exchange	Chemicals secreted by endocrine glands
Process of ingestion in the ameba	Organ that pumps blood throughout an earthworm's body	Most toxic nitrogenous waste	Organism whose skin acts as the primary organ of respiration	Three major regions of a nerve cell
End product of chemical digestion of protein	Intracellular circulation of cytoplasm	Insoluble crystals of nitrogenous waste	Ethyl alcohol or lactic acid	Two organisms possessing a ventral nerve cord

HUMAN PHYSIOLOGY: DOUBLE JEOPARDY

Regulation	Circulation	Excretion	Digestion	Respiration
Two systems responsible for regulation	Thick-walled vessels carrying blood from the heart	One of the two major functions of the kidney	Of glucose, starch and protein, the one that does not require digestion	Functional units for gas exchange within the lung
Three major divisions of the brain	Water containing dissolved salts proteins, nutrients and wastes, among other things	Although some excretory products are removed by this process, its primary function is temperature regulation	Organ whose protein digesting enzymes function best in an acidic pH	Gas which regulates breathing rate
One of the hormones secreted by the pituitary gland	One specific type of white blood cell	Microscopic units of filtration within the kidney	Peristalsis	Directions in which diaphragm and rib cage move during inhalation
Disease resulting from insufficient insulin production	One method to acquire immunity	Path of urea from kidney to external environment	Primary organ of chemical digestion and absorption	Other than an opening to the air, three respiratory functions of nasal passages
Receptor → sensory neurons, → motor neuron → effector	Chamber of the heart receiving blood from the lungs	Two of its functions include destruction of worn out red blood cells and deamination of amino acids	Specific function of bile	Structure supported by cartilagenous rings and ciliated mucus membrane

ANSWERS TO SAMPLE JEOPARDY GAMES

Animal Maintenance: Single Jeopardy	Human Physiology: Double Jeopardy

Nutrition:
100 What is a heterotroph?
200 What is chemical digestion?
300 What is a Hydra (Coelenterate)?
400 What is phagocytosis?
500 What are amino acids?

Transport:
100 What is active transport?
200 What is a grasshopper (insect)
300 What is the pigment that carries oxygen in the red blood cells?
400 What are aortic arches?
500 What is cyclosis?

Excretion:
100 What are metabolic wastes
200 What is moist skin?
300 What are nephrons?
400 What is ammonia?
500 What is uric acid?

Respiration:
100 What is the equation for aerobic respiration?
200 What is a respiratory membrane?
300 What are tracheal tubes?
400 What is an earthworm? (Annelid)
500 What is an end product of anaerobic respiration?

Regulation:
100 What is a neuron?
200 What is a Hydra (Coelenterate)?
300 What are hormones?
400 What are dendrites, axon and cyton (cell body)
500 What are earthworm and grasshopper?

Regulation:
200 What are the nervous and endocrine systems?
400 What are the cerebrum, cerebellum and medulla?
600 What is growth hormone (or FSH, ACTH, etc.)
800 What is diabetes?
1000 What is the path of a reflex arc?

Circulation:
200 What are arteries?
400 What is plasma?
600 What is a lymphocyte (or phagocyte, and so on)
800 What is inborn (or passive, active or maternal)?
1000 What is the left atrium?

Excretion:
200 What is filtration or maintenance of fluid balance?
400 What is perspiration?
600 What are nephrons?
800 What is ureter to bladder to urethra?
1000 What is the liver?

Digestion:
200 What is glucose?
400 What is the stomach?
600 What are wavelike contractions that push food through the digestive tract?
800 What is the small intestine?
1000 What is to emulsify fats?

Respiration:
200 What are alveoli?
400 What is carbon dioxide
600 What is down for diaphragm and up & out for ribs?
800 What is warming, filtering, and moistening the air?
1000 What is the trachea?

BIO BINGO

INTRODUCTION: This activity is a game of Bingo, but with a bit of a twist, as you shall see in the *Procedure* section. The objective of this activity is to provide you with an interesting method for studying and review. Use it well, and have fun.

PURPOSE:

- To answer BINGO questions correctly so that you are the first to have five answers in a row; horizontally, diagonally or vertically.

PROCEDURE:

1. Obtain a Bio Bingo game board and tokens.

2. As the moderator reads a question in one of the five categories, look over your game board to see if the correct answer is present. If it is, cover the answer with a token.

3. Continue in this fashion until someone has covered five answers in a straight line; horizontally, vertically or diagonally. When this occurs, the winning student should call out *"Bingo."*

4. Have the moderator check the winning game board, to be sure that all the covered answers are actually the correct ones for the questions read.

5. If the board is, indeed, a winner, the game is over. If one or more answers are wrong, however, the game continues with the remaining players. The individual who erroneously called out *"Bingo"* is disqualified.

   The categories in Bio Bingo are as follows:

   B = biochemistry

   I = interdependence

   N = notables

   G = genetics

   O = organ systems

SUGGESTIONS FOR FURTHER STUDY:

- Make up your own questions for Bio Bingo. Play with other members of the class. Exchange game questions.

- Make up your own categories for Bio Bingo questions. What other topics might begin with the letters available? Again, play with other members of the class; exchange game categories and questions.

BLANK GAME BOARD
BIOLOGY BINGO

B Biochemistry	I Interdependence	N Notables	G Genetics	O Organs
		FREE		

SAMPLE BINGO QUESTIONS

B(Biochemistry) **I**(Interdependence) **N**(Notables) **G**(Genetics) **O**(Organ Systems)

BIOCHEMISTRY

What is the name of the chemical process by which simple molecules are combined to form more complex molecules?

What is the basic building block of starch?

What is the name of the bond that joins amino acids within a protein?

What element is found in all organic molecules?

What chemicals serve as organic catalysts?

What is the substrate upon which a protease works?

What class of organic compounds is formed from the synthesis of many nucleotides?

Adenine, thymine, cytosine and guanine are found in what compound?

Which organic compounds serve as the primary source of energy in living systems?

Which organic molecules are composed of glycerol and fatty acids?

INTERDEPENDENCE

What is the name of the symbiotic relationship in which one organism benefits and another is harmed?

What organisms make–up the lowest level of a food chain?

What do you call organisms that eat only animals?

What is the term given to the intricate energy relationships among plants and animals in a community?

What is the primary energy source for all organisms on Earth?

What are the non–living factors in an ecosystem called?

What type of organism recycles dead matter back into the food chain?

What type of symbiosis is exhibited by lichen?

What term describes a consumer that only eats producers?

What biochemical process produces the carbon dioxide that is needed by photosynthetic plants?

NOTABLES

Who discovered the path of blood circulation?

Who is considered to be the "Father of Genetics?"

Name one scientist that helped support the theory of random synthesis and the chemical evolution of life.

Who helped bring the downfall of the theory of spontaneous generation, as well as demonstrate the microbial nature of disease?

Who is credited with inventing the microscope?

Name one of the scientists responsible for the cell theory.

Who developed the basis for the modern system of classification?

Name one of the scientists responsible for determing the structure of DNA.

Who is considered the "Father of Evolutionary Theory?"

Whose experiments most clearly demonstrated the nature of conditioning?

GENETICS

What is the term given to a section of DNA responsible for the production of a specific protein?

How many pairs of chromosomes are found in human diploid cells?

What is the term that refers to a trait that is always expressed if it is present in the genes?

What term refers to the physical expression of a specific genotype?

If two heterozygous tall pea plants are crossed, what percentage of the offspring will be tall?

What is the term that refers to traits carred on the X chromosome?

What type of inheritance is illustrated by pink flowers that are produced from a cross between a red flowering plant and a white flowering plant?

The genetic code of DNA is translated into a specific protein on what cell organelle?

What chemical is responsible for transferring the DNA genetic code to the cytoplasm of a cell?

What type of mutation occurs when a pair of homologous chromosomes fail to separate during meiosis?

ORGAN SYSTEMS

What blood vessels connect arteries and veins?

In what organ is fat digested?

What are the structures that prevent blood from flowing in the wrong direction through the circulatory system?

Name the part of the brain responsible for voluntary action.

What are the functional units of the kidney?

What tissue connects bone to muscle?

Where is carbon dioxide removed from the bloodstream?

What two systems are responsible for regulation of the body?

In which gland is insulin produced?

What organ is responsible for the production of sperm cells?

SAMPLE GAME BOARD

B	I	N	G	O
DNA	parasitism	Pasteur	75%	capillaries
peptide	herbivore	Leewenhoek	m-RNA	cerebrum
lipid	food web	**FREE**	dominant	small intestine
carbon	mutualism	Crick	gene	alveoli
protein	sunlight	Darwin	sex linked	testes

LIGHTS, CAMERA, ACTION!
PERFORM YOUR OWN BIOLOGY SHOW

INTRODUCTION: This activity is an opportunity for you to put your creative talents to work. Using any or all of the information you learned this year in biology, create, produce, and perform a show! The type of show you create is entirely up to you and the other members of your group. The only criteria are that your show contains accurate scientific information and that you have some fun in the process.

Take–offs on soap operas, game shows, and T.V. situation comedies work quite well ("High School Hospital" or "What's My Function," for example). Or, you may want to put together your own musical review, writing biology songs ("If I Only Had A Cerebrum"). Stand–up comics may want to get into the act with original biology jokes. Your production should be in good taste, and contain biological facts, concepts or questions. Hollywood, here you come!

PURPOSE:

- How may biology be used as the basis for an entertainment vehicle?

MATERIALS:

Assorted props that you select
Creativity
Imagination
Knowledge of biology

PROCEDURE:

With the help of your teacher, divide yourselves into groups of 4–8 students. All members of the group must be able to work together in the spirit of cooperation. Your first assignment is to give your group a name (You've all heard of the famous female singing group the "Bioettes," haven't you?)

Once your group is formed, discuss various possibilities for your production. Once you have agreed upon a format, you must write your script. Depending upon the number, talent and interest of the group's members, you may choose to delegate different responsibilities to different members, or you may all want to work together. Writing your script is only one aspect of this activity. Research, song writing, acting, directing, etc., are also important components. Not all members of the group need be actors, but everyone **must** be involved in some aspect of the production.

If your school has a drama or speech department or club, you may want to work with teachers in these areas. They can often offer valuable assistance and suggestions.

Your teacher will be available for assistance in technical matters. Any questions of scientific accuracy should be well researched. Your teacher will also tell you the date and manner of presentation for your "show." You will be performing in front of other members of your class, or, possibly, before other students in the school, your family, or even the general public. Your teacher will discuss this with the class.

Remember, imagination, humor or intrigue may be important for a successful production, but scientific accuracy is essential. This activity should be a lot of fun, so enjoy yourselves. One or more prizes (to be determined by your teacher) will be awarded at the conclusion of all presentations.